PROBABILITY THEORY
for
Actuarial Mathematics

Jai Heui KIM

Kyowoo Publishing Company

About the Author

Jai Heui Kim, Professor of Mathematics at Pusan National University (Korea) since 1986, graduated from this university in 1979. He received his doctor of science degree from Osaka University (Japan) in 1986 for a thesis entitled "Stochastic calculus related to non-symmetric Dirichlet forms".

Kyowoo Publishing Company
148-19 Jaekidong, Dongdaemoonku,
Seoul 132-062, Korea

Phone : 82-2-925-2861
Fax : 82-2-925-2860
e-mail : kw2861@chollian.net
http://www.kyowoo.co.kr

ISBN 978-89-8172-667-6 (93310)
Printed in the Republic of korea.

Preface

Probability theory is a branch of mathematics that finds applications in every area of scholarly activity from music to physics, and in daily experience from weather prediction to predicting the risks on a financial market.

This book is designed to give undergraduate and graduate students a through understanding of the concepts of probability theory and its applications to the insurance models, and in particular, to help in preparation for the Society of Actuaries/Casualty Actuarial Society Exam P/1. The only formal prerequisite is calculus, but an additional degree of mathematical maturity may be helpful.

This book consists of two parts. The first part consists Chapter 1 through 5, which can be used for a one-semester course in probability. The second part discusses 106 questions which are cited from home site on the World Wide Web, http://www.soa.org([4]). This part emphasizes to classify these questions by subjects and give more heuristic solutions, than original ones, in which mathematical meaning for given question, assumptions that we are given, key tool to solve the question and detail calculation can be found.

I would like to thank the many students who read the manuscript. I wish to thank the publishers of SOA/CAS for permission to include 106 questions from [4] in this book. I shall be grateful to any reader who finds and brings some misprints, errors and ambiguities of presentation. Finally I acknowledge with thanks my indebtedness to Head P.K. Oh of Kyowoo Publishing Co. for his cooperation in this work.

J.H. Kim

February 25, 2007

Contents

Part 1

Probability Theory

CHAPTER 1

Introduction to Probability

1.1. Sample spaces and events

The collection or set of all possible distinct(that is, collectively exhaustive and mutually exclusive) outcomes of an experiment is called the **sample space** of the experiment and is denoted by the symbol S. For the present an experimental outcome is considered to be a single outcome, not a set of outcomes. We call a single outcome a **sample point**. A subset of outcomes will be defined later as an event.
A sample space is called **discrete** if it is a finite or countably infinite set. It is called **continuous** otherwise.
Some examples of sample spaces are following.

Example 1.1.1. Suppose a ball is selected from an urn containing three identical balls, labeled 1,2, and 3. The urn is first shaken to randomize the position of the balls, and a ball is selected. Then we get a sample space

$$S_1 = \{1, 2, 3\}.$$

Example 1.1.2. If the experiment consists of tossing two dice, then we get the following discrete sample space

$$S_2 = \{(i,j)|\ i,j = 1,2,3,4,5,6\}.$$

where the outcome (i,j) is said to occur if i appears on the leftmost die and j on the other die.

Example 1.1.3. Three students x, y and z, in order, toss a fair die. The first one to toss an even number wins. The game continuous until someone tosses an even number. Let s_n be an elementary outcome that n-th toss is the first even number, i.e., it is an outcome that j-th tosses, $j = 1, 2, \cdots, n-1$ are odd and n-th toss is even. Then the sample space of this game is given by

$$S_3 = \{s_i|\ i = 1, 2, 3, \cdots\}.$$

Example 1.1.4. If the experiment consists of measuring the lifetime of a given computer memory chip in specified environment, then we have the sample space S_4 consisting of all nonnegative real numbers. That is,

$$S_4 = \{x|\ 0 \le x < \infty\}.$$

Example 1.1.5. If the experiment consists of picking two numbers x and y at random between 0 and 1, then the sample space is the unit square in $\mathbf{R}^2$. That is,

$$S_5 = \{(x,y) \in \mathbf{R}^2|\ 0 \le x, y \le 1\}.$$

S_1, S_2 and S_3 are discrete sample spaces and S_4 and S_5 are continuous. Any collection of sample points, or any subset of the sample space is referred to as an **event**. When the experiment is performed, and if the observed outcome is contained in the event E, then we say that the event E **occurred**.

Example 1.1.6. (1) In Example 1.1.1, $E = \{1, 3\}$ is the event that an odd-numbered ball is selected.
(2) In Example 1.1.2, $E = \{(3, 6), (4, 5), (5, 4), (6, 3)\}$ is a subset of S_2, which is the event that the sum of the dice equals 9.
(3) In Example 1.1.3, if $E = \{s_{3k-1} | \ k = 1, 2, \cdots\}$, then E is the event that the second student y will win in the game.
(4) In Example 1.1.4, $E = \{t : 800 < t < 1200\}$ is the event that the computer memory chip lasts more than 800 hours but less than 1200 hours.
(5) In Example 1.1.5, $E = \{(x, y) \in \mathbf{R}^2 | \ 0 \leq x \leq 1, \frac{1}{2} \leq y \leq 1\}$ is the event that a point in the upper half region of S_5 is picked.

We can combine events using set operations to obtain other events. We can also express complicated events as combinations of sample points.

The **union of two events** A and B is denoted by $A \cup B$ and is defined as the set of outcomes that are either in A or in B or both. The event $A \cup B$ occurs if ether A or B occurs.

The **intersection of two events** A and B is denoted by $A \cap B$ and is defined as the set of outcomes that are in both A and B. The event $A \cap B$ occurs when both A and B occur. Two events are said to be **mutually exclusive** if their intersection is null event ϕ. Mutually exclusive events cannot occur simultaneously.

The **complement of an event** A is denoted by A^c and is defined as the set of all outcomes not in A. The event A^c occurs when the event A does not occur and vice versa.

The union and intersection operations can be repeated for an arbitrary number of events. Thus the event

$$\bigcup_{k=1}^{n} A_k = A_1 \cup A_2 \cup \cdots \cup A_n$$

occurs if one or more of the events A_k occur. The event

$$\bigcap_{k=1}^{n} A_k = A_1 \cap A_2 \cap \cdots \cap A_n$$

occurs if all of the events $A_1, A_2, \cdots, A_n$ occur. The operations can also be applied to a countably infinite sequence of events $A_1, A_2, \cdots$.

We say that events $B_1, B_2, \cdots, B_n$ form a **partition of event** A if $A = \cup_{i=1}^{n} B_i$ and the B_i's are mutually exclusive.

1.2. The axioms of probability

Suppose that the experiment of Example 1.1.1 is repeated n times under identical conditions. Let $N_k(n)$, $k = 1, 2, 3$, be the number of times in which the outcome is ball k. The **relative frequency** of outcome k is defined by

$$f_k(n) = \frac{N_k(n)}{n}.$$

Then the relative frequency $f_k(n)$ has the following properties.

$$0 \leq f_k(n) \leq 1, \quad \sum_{k=1}^{3} f_k(n) = \frac{1}{n} \sum_{k=1}^{3} N_k(n) = 1.$$

It is common experience that the relative frequency of the occurrence of a random event during repeated trials tend to a recognizable limit as the number of trials become large. This phenomenon is called **statistical regularity**, and the limiting value of the relative frequency is the probability of the event in question.

In our example, we have

$$\lim_{n\to\infty} f_k(n) = p_k. \tag{1.2.1}$$

The constant p_k is called the **probability** of the outcome k. The equality (1.2.1) is referred to as **Bernoulli's Law of Large Numbers** which will be discussed late.

For any event A, define $P(A) = \sum_{k\in A} p_k$. Then we can see easily that $0 \leq P(A) \leq 1$ and $P(S_1) = 1$. Moreover, by the fact that the relative frequency $f_A(n)$ of the event A in the n repetitions of the experiment satisfies the property $f_A(n) = \sum_{k\in A} f_k(n)$, we have

$$P(A \cup B) = P(A) + P(B) \quad \text{if} A \cap B = \emptyset.$$

A *probability measure* for a random experiment is a rule that assigns probability to the events of the experiment. Thus a probability measure is a set function that assigns a number to events. In the example above, we found some properties of relative frequency that any definition of probability should satisfy. The axioms of probability formally state that a probability measure must satisfy these properties. From now we develop a number of results that follow this set of axioms.

Definition 1.2.1. Consider an experiment with sample space S. A **probability measure** P for the experiment is a set function that assigns to any event A a number $P(A)$, called the **probability** of A, that satisfies the following axioms:

Axiom I $0 \leq P(A) \leq 1$

Axiom II $P(S) = 1$

Axiom III If $A \cap B = \emptyset$, then $P(A \cup B) = P(A) + P(B)$

Axiom III' If $A_1, A_2, \cdots$ are mutually exclusive events, then

$$P\left(\bigcup_{k=1}^{\infty} A_k\right) = \sum_{k=1}^{\infty} P(A_k).$$

Axiom I, II and III are enough to deal with experiments with finite sample spaces. In other to handle experiments with infinite sample spaces, Axiom III needs to be replaced by Axiom III'. Note that Axiom III' includes Axiom III as a special case, by letting $A_k = \emptyset$ for $k \geq 3$. Thus we really only need Axiom I, II and III'.

Theorem 1.2.2. *It holds* $P(\emptyset) = 0$.

PROOF. We may write, for any event A, $A = A \cup \emptyset$. Since A and $\emptyset$ are mutually exclusive, it follows from Axiom III that $P(A) = P(A \cup \emptyset) = P(A) + P(\emptyset)$. From this the conclusion of the theorem is immediate. □

Theorem 1.2.3. *For any event* A, $P(A) = 1 - P(A^c)$.

PROOF. We may write $S = A \cup A^c$ and, using Axiom II and III, we obtain $1 = P(S) = P(A) + P(A^c)$. □

Theorem 1.2.4. *If* A *and* B *are any two events, then*

$$P(A \cup B) = P(A) + P(B) - P(A \cap B).$$

PROOF. The idea of this proof is to decompose $A \cup B$ and B into mutually exclusive events and then apply Axiom III. We see that

$$A \cup B = A \cup (B \cap A^c),$$
$$B = (A \cap B) \cup (B \cap A^c).$$

Hence

$$P(A \cup B) = P(A) + P(B \cap A^c),$$
$$P(B) = P(A \cap B) + P(B \cap A^c).$$

Subtracting the second equality from the first yields

$$P(A \cup B) - P(B) = P(A) - P(A \cap B)$$

and hence the results follows. □

Remark 1.2.5. Let A, B and C be any three events. Writing $A \cup B \cup C$ as $(A \cup B) \cup C$ and applying the result of the above theorem, we have

$$P(A \cup B \cup C) = P(A) + P(B) + P(C) - P(A \cap B) - P(A \cap C) \\ - P(B \cap C) + P(A \cap B \cap C).$$

Moreover, if $A_1, A_2, \cdots, A_k$ are any k events, by using mathematical induction, then we have

$$\begin{aligned} P(A_1 \cup A_2 \cup \cdots \cup A_k) &= \sum_{i=1}^{k} P(A_i) - \sum_{i<j=2}^{k} P(A_i \cap A_j) \\ &\quad + \sum_{i<j<m=3}^{k} P(A_i \cap A_j \cap A_m) \\ &\quad + \cdots + (-1)^{k-1} P(A_1 \cap A_2 \cap \cdots \cap A_k). \end{aligned} \tag{1.2.2}$$

Theorem 1.2.6. *If $A \subset B$, then $P(A) \le P(B)$.*

PROOF. We may decompose B into mutually exclusive events as follows: $B = A \cup (B \cap A^c)$. Hence $P(B) = P(A) + P(B \cap A^c) \ge P(A)$, since $P(B \cap A^c) \ge 0$ from Axiom I. □

Theorem 1.2.7. *If $A_1, A_2, \cdots$ are any sequence of events, then it holds*

$$P\left(\bigcup_{k=1}^{\infty} A_k\right) \le \sum_{k=1}^{\infty} P(A_k).$$

PROOF. Define a sequence $B_1, B_2, \cdots$ of events by

$$\begin{aligned} B_1 &= A_1 \\ B_2 &= A_1 \cup A_2 - A_1 \\ B_k &= \bigcup_{i=1}^{k} A_i - \bigcup_{i=1}^{k-1} A_i, \quad k = 3, 4, \cdots. \end{aligned}$$

Then $B_1, B_2, \cdots$ are pairwise mutually exclusive, $B_k \subset A_k$ and $\cup_{k=1}^{\infty} B_k = \cup_{k=1}^{\infty} A_k$. Thus it holds

$$P\left(\bigcup_{k=1}^{\infty} A_k\right) = P\left(\bigcup_{k=1}^{\infty} B_k\right) = \sum_{k=1}^{\infty} P(B_k) \leq \sum_{k=1}^{\infty} P(A_k).$$

□

Example 1.2.8. Consider Example 1.1.1. Since all distinct sample points are mutually exclusive, by Axiom III, the probability of the event $E = \{1, 3\}$ is given by $P(E) = P(\{1, 3\}) = P(\{1\}) + P(\{3\})$. If the outcomes $1, 2, 3$ in the sample space S_1 are **equally likely**, that is,

$$P(\{1\}) = P(\{2\}) = P(\{3\}) = \frac{1}{3},$$

then we have $P(E) = 1/3 + 1/3 = 2/3$.

In the above example, '2/3' means the number of outcomes in the event E divided by the total number of outcomes in the sample space S_1. In general, *if outcomes are equally likely, then the probability of an event is equal to the number of outcomes in the event divided by the total number of outcomes in the sample space.* Next section discusses counting methods that are useful in finding probabilities in experiments that have equally likely outcomes.

From now, we denote $P(\{x\})$ by $P(x)$ for any sample point x.

Example 1.2.9. In Example 1.1.2, we assume that two dice are fair. Then all 36 outcomes are equally likely, i.e., the probability of any sample point is 1/36. If $E = \{(3, 6), (4, 5), (5, 4), (6, 3)\}$, then

$$P(E) = P((3, 6)) + P((4, 5)) + P((5, 4)) + P((6, 3)) = \frac{4}{36} = \frac{1}{9}.$$

Example 1.2.10. In Example 1.1.3, if $E = \{s_{3k-1} |\ k = 1, 2, \cdots\}$, then E is the event that the second student y will win in the game. Find $P(E)$.

SOLUTION. If we denote by A_i the event that i-th throw is even, then $P(A_i) = P(A_i^c) = 1/2$ for all $i = 1, 2, \cdots$ and so the probability of s_n is given by

$$\begin{aligned}
P(s_n) &= P(A_1^c \cap A_2^c \cap \cdots A_{n-1}^c \cap A_n) \\
&= P(A_1^c) \cdot P(A_2^c) \cdots P(A_{n-1}^c) \cdot P(A_n) \\
&= \left(\frac{1}{2}\right)^{n-1} \left(\frac{1}{2}\right) \\
&= \frac{1}{2^n}.
\end{aligned}$$

Thus we have

$$\begin{aligned}
P(E) &= P(\{s_2, s_5, s_8, \cdots\}) \\
&= P(s_2) + P(s_5) + P(s_8) + \cdots \\
&= \frac{1}{2^2} + \frac{1}{2^5} + \frac{1}{2^8} + \cdots \\
&= \frac{1}{2^2} \left\{1 + \frac{1}{8} + \frac{1}{8^2} + \cdots \right\} \\
&= \frac{1}{4} \left\{\frac{1}{1 - 1/8}\right\} \\
&= \frac{2}{7}.
\end{aligned}$$

Example 1.2.11. In Example 1.1.4, the sample space was $S = [0, \infty)$. Define a set function P for any interval $(s, t), 0 \le s \le t \le \infty$, by

$$P((s,t)) = e^{-s\alpha} - e^{-t\alpha},$$

where $\alpha > 0$. Then we can see easily that P satisfies Axiom I,II and III' and so it is a probability measure for this experiment.

Example 1.2.12. In Example 1.1.5, the sample space was a unique square. If we assume that all pairs of numbers in this unit square are equally likely to be selected, then it is reasonable to define the probability of any region E inside the unique square by the area of E. For example, let $A = \{(x,y)|\ 0 \le x \le 1, 1/2 \le y \le 1\}$ and $B = \{(x,y)|\ 0 \le x \le 1, 0 \le y \le x\}$. Clearly each of these regions has area 1/2. Thus we have $P(A) = P(B) = 1/2$.

1.3. Combinatorial principles - permutations and combinations

We will shortly see that sometimes the numbers we calculate for the number of elements in an event may become quite large. Furthermore, they are often related to one another in a very special way, such as in permutations and combinations of elements in a set. In this section we will define these terms shortly. To help us make such calculations it is convenient to have a special notation, called factorial.

Definition 1.3.1. Let n be any positive integer. We define $n! = n(n-1)(n-2)\cdots 2\cdot 1 = \prod_{k=1}^{n} k$ and call it n **factorial**. We also define $0! = 1$.

Given n distinct object, without replacement, the number of different ways in which the objects may be ordered(permuted) is $n!$. For example, the set of 3 letters $\{A, B, C\}$ can be ordered in $3! = 6$ ways. All permutations are ABC, ACB, BAC, BCA, CAB, CBA.

Definition 1.3.2. A **permutation** of size k from a set of size n is an ordered subset of size k. We denote the number of different permutations of size k from a set of size n, without replacement, by ${}_nP_k$, i.e.,

$${}_nP_k = n(n-1)(n-2)\cdots(n-k+1) = \frac{n!}{(n-k)!}.$$

Using the set {A,B,C} again, the number of different permutations of size 2 is ${}_3P_2 = 3!/(3-2)! = 6/1 = 6$. These are AB, AC, BA, BC, CA, CB.

Given n objects, of which n_1 are of Type 1, n_2 are of Type 2,$\cdots$, and n_p are of Type p ($p \geq 1$ is an integer), and $n = n_1 + n_2 + \cdots + n_p$, the number of ways of ordering all n objects is $n!/(n_1!n_2!\cdots n_p!)$, which is sometimes denoted by $\binom{n}{n_1 n_2 \cdots n_p}$ and is called a multinomial coefficient.

Definition 1.3.3. A **combination** of size k from a set of size n is an unordered subset of size k. We denote the number of different combinations of size k from a set of size n, without replacement, by $\binom{n}{k}$ or ${}_nC_k$, i.e.,

$$\binom{n}{k} = \frac{{}_nP_k}{k!} = \frac{n!}{k!(n-k)!}\left(= \binom{n}{n-k}\right),$$

which is called a **binomial coefficient** and is read "n choose k" or "n taken k at a time".

Notice the following elementary properties of the binomial coefficients.

$$\binom{n}{k} = \frac{n-k+1}{k}\binom{n}{k-1},$$

$$\binom{n}{k} = \frac{n}{k}\binom{n-1}{k-1}$$

$$(a+b)^n = \sum_{k=0}^{n}\binom{n}{k}a^k b^{n-k} \tag{1.3.1}$$

and

$$(x_1 + x_2 + \cdots + x_p)^n = \sum_{k_1+k_2+\cdots+k_p=n}\binom{n}{k_1 k_2 \quad k_p}x_1^{k_1}x_2^{k_2}\cdots x_p^{k_p}. \tag{1.3.2}$$

The equalities (1.3.1) and (1.3.2) are known as the binomial theorem and the multinomial theorem, respectively.

Using the set $\{A, B, C\}$ again, the number of different combinations of size 2 is $\binom{3}{2} = {}_3P_2/2! = 3!/\{2!(3-2)!\} = 3$. These are $\{A, B\}$, $\{A, C\}$, and $\{B, C\}$.

Example 1.3.4. Let S be the set of all permutations of 6 objects. Find the probability of the set of all permutations with at least one fixed point.

SOLUTION. Let A_i be the set of all permutations that the i-th object is unchanged. Then the set $\cup_{i=1}^{6} A_i$ is the set of all permutations with at least one fixed point. It is sufficient to find $P(\cup_{i=1}^{6} A_i)$. Since

$$P(A_i) = \frac{5!}{6!}, \qquad i = 1, 2, 3, 4, 5, 6,$$

we have

$$\sum_{i=1}^{6} P(A_i) = \binom{6}{1}\frac{5!}{6!} = \frac{6!5!}{1!5!6!} = 1.$$

And since

$$P(A_i \cap A_j) = \frac{4!}{6!}, \qquad i < j, \quad i, j = 1, 2, 3, 4, 5, 6,$$

we have

$$\sum_{i<j}^{6} P(A_i \cap A_j) = \binom{6}{2}\frac{4!}{6!} = \frac{6!4!}{2!4!6!} = \frac{1}{2!}.$$

By the similar calculation, we have

$$\sum_{i<j<k}^{6} P(A_i \cap A_j \cap A_k) = \frac{1}{3!}$$

and so on. Hence, by the equality (1.2.2), we have

$$\begin{aligned} P\left(\bigcup_{i=1}^{6} A_i\right) &= 1 - \frac{1}{2!} + \frac{1}{3!} - \frac{1}{4!} + \frac{1}{5!} - \frac{1}{6!} \\ &= 1 - \frac{1}{2} + \frac{1}{6} - \frac{1}{24} + \frac{1}{120} - \frac{1}{720} \\ &= \frac{441}{720} = \frac{49}{80}. \end{aligned}$$

Example 1.3.5. In a game of bridge the probability that a hand of 13 cards contains 5 spades, 3 hearts, 4 diamonds, and 1 club is

$$\frac{\binom{13}{5}\binom{13}{3}\binom{13}{4}\binom{13}{1}}{\binom{52}{13}}.$$

1.4. Conditional probability and independence

In this section we introduce two of the most important concepts in probability theory, which are conditional probability and independence of events.

Definition 1.4.1. Let B be an event with $P(B) > 0$. For arbitrary event A we define

$$P(A|B) = \frac{P(A \cap B)}{P(B)}$$

and call it the **conditional probability** of event A given that event B has occurred.

Example 1.4.2. Consider a hand of five cards in a game of poker. If the cards are dealt at random, there are $\binom{52}{5}$ possible hands of five cards each. Let A and B be the events that at least 3 cards are spades and that all five cards are spades, respectively. Let A_k, $k = 1, 2, 3, 4, 5$, be the event that a hand of five cards contains k spades. Then

$$\begin{aligned} P(A) &= P(A_3 \cup A_4 \cup A_5) \\ &= \frac{\binom{13}{3}\binom{39}{2} + \binom{13}{4}\binom{39}{1} + \binom{13}{5}}{\binom{52}{5}}. \end{aligned}$$

And we have clearly

$$P(A \cap B) = P(B) = \frac{\binom{13}{5}}{\binom{52}{5}}.$$

Thus it follows that

$$\begin{aligned} P(B|A) &= \frac{P(A \cap B)}{P(A)} \\ &= \frac{\binom{13}{5} / \binom{52}{5}}{\left\{ \binom{13}{3}\binom{39}{2} + \binom{13}{4}\binom{39}{1} + \binom{13}{5} \right\} / \binom{52}{5}}. \end{aligned}$$

Example 1.4.3. Suppose that an urn contains 7 black balls and 5 white balls. We draw 2 balls from the urn without replacement. If we assume that at each draw each ball in the urn is equally likely to be chosen, what is the probability that both balls drown are black?

SOLUTION. Let B_1 and B_2 denote the events that the first and second balls drown are black, respectively. If the first ball selected is black, then there are 6 remaining black balls and 5 white balls. So we have $P(B_2|B_1) = 6/11$. Since $P(B_1) = 7/12$, the desire probability is

$$\begin{aligned} P(B_1 \cap B_2) &= P(B_2|B_1)P(B_1) \\ &= \frac{6}{11} \cdot \frac{7}{12} \\ &= \frac{7}{22}. \end{aligned}$$

Alternatively, this probability can be computed by

$$P(B_1 \cap B_2) = \frac{\binom{7}{2}}{\binom{12}{2}} = \frac{7}{22}.$$

Theorem 1.4.4 (Bayes' rule). *Let events $B_1, B_2, \cdots, B_n$ form a partition of a sample space S. Then for any event A with $P(A) > 0$,*

$$P(B_k|A) = \frac{P(A|B_k)P(B_k)}{\sum_{i=1}^{n} P(A|B_i)P(B_i)}, \quad k = 1, 2, \cdots, n.$$

PROOF. Since

$$P(A \cap B_k) = P(A|B_k)P(B_k)$$

and

$$P(A) = \sum_{i=1}^{n} P(A \cap B_i) = \sum_{i=1}^{n} P(A|B_i)P(B_i), \tag{1.4.1}$$

it follows that

$$P(B_k|A) = \frac{P(A \cap B_k)}{P(A)} = \frac{P(A|B_k)P(B_k)}{\sum_{i=1}^{n} P(A|B_i)P(B_i)}.$$

□

The equality (1.4.1) is known as the **theorem of total probability**. As a special case, for any event A and B, it holds that

$$P(A) = P(A|B)P(B) + P(A|B^c)P(B^c). \tag{1.4.2}$$

Example 1.4.5. An insurance company believes that people can be divided into two classes: those who are accident prone and those who are not. Their statistics show that an accident-prone person will have an accident at some time within a fixed one year period with probability 0.6, whereas this probability is 0.3 for a non-accident-prone person. Assume that 20 percent of the population is accident prone.
(i) Find the probability that a new policyholder will have an accident within a year of purchasing a policy.
(ii) Suppose that a new policyholder has an accident within a year of purchasing a policy. Find the probability that he or she is accident prone.

SOLUTION. Let A denote the event that the policyholder will have an accident within a year of purchasing a policy and B denote the event that the policyholder is accident prone.
(i) Using the equality (1.4.2), the desired probability $P(A)$ is given by

$$\begin{aligned} P(A) &= P(A|B)P(B) + P(A|B^c)P(B^c) \\ &= 0.6 \times 0.2 + 0.3 \times 0.8 \\ &= 0.4. \end{aligned}$$

(ii) The desired probability is $P(B|A)$. From (i), it follows that

$$\begin{aligned} P(B|A) &= \frac{P(A \cap B)}{P(A)} \\ &= \frac{P(A|B)P(B)}{P(A)} \\ &= \frac{0.6 \times 0.2}{0.4} = \frac{3}{10} \\ &= 0.3. \end{aligned}$$

Example 1.4.6. Urn I contains one white and two black marbles, urn II contains one black and two white marbles, and urn III contains three black and three white marbles. A die is rolled. If 1,2, or 3 shows up, urn I is selected; if 4 shows up, urn II is selected; if 5 or 6 shows up, urn III is selected. A marble is drawn at random from the selected urn. What is the conditional probability that urn II is selected given that a white marble has been drawn?

SOLUTION. Let H be the event that the marble drawn is white. And let U, V, W, respectively, denote the events that the urn selected is I, II, III. Then the desired conditional probability is $P(V|H)$. From Bayes' rule, we have

$$
\begin{aligned}
P(V|H) &= \frac{P(H|V)P(V)}{P(H|U)P(U)+P(H|V)P(V)+P(H|W)P(W)} \\
&= \frac{(2/3)\cdot(1/6)}{(1/3)\cdot(3/6)+(2/3)\cdot(1/6)+(3/6)\cdot(2/6)} \\
&= \frac{1/9}{4/9} = \frac{1}{4}.
\end{aligned}
$$

Definition 1.4.7. The two events A and B are said to be **independent** if and only if

$$P(A \cap B) = P(A)P(B).$$

By the conditional probability we see that if $P(B) > 0$, then A and B are independent if and only if $P(A|B) = P(A)$. Thus, two events with positive probability are independent if knowing the outcome of one event does not affect the probability of the other event.

Example 1.4.8. A coin is tossed twice. Let A be the event that the first toss is a head and B the event the two outcome are the same. Then A and B are independent. In fact, the sample space is

$$S = \{HH, HT, TH, TT\}$$

where, for example, HT is an elementary outcome "heads on the first toss and tails on the second". Clearly $A = \{HH, HT\}$ and $B = \{HH, TT\}$. Since $A \cap B = \{HH\}$, we can see easily

$$P(A \cap B) = \frac{1}{4} = \frac{1}{2} \cdot \frac{1}{2} = P(A)P(B)$$

and so A and B are independent events.

Alternatively, we have

$$P(A|B) = \frac{P(A \cap B)}{P(B)} = \frac{1/4}{1/2} = \frac{1}{2} = P(A).$$

Thus, A and B are independent.

Definition 1.4.9. We say that three event A, B and C are **mutually independent** if and only if all the following conditions hold:

$$\begin{aligned} P(A \cap B) &= P(A)P(B), \\ P(A \cap C) &= P(A)P(C), \\ P(B \cap C) &= P(B)P(C), \\ P(A \cap B \cap C) &= P(A)P(B)P(C). \end{aligned}$$

We finally generalize this notion to n events in the following definition.

Definition 1.4.10. The n events $A_1, A_2, \cdots, A_n$ is said to be mutually independent if and only if, for every subfamily $\{A_{i_1}, A_{i_2}, \cdots, A_{i_k}\}$, $1 \le k \le n$, of this family, the following relation holds:

$$P(A_{i_1} \cap A_{i_2} \cap \cdots \cap A_{i_k}) = P(A_{i_1})P(A_{i_2}) \cdots P(A_{i_k}).$$

Example 1.4.11. Assume that we toss two dice. Define the event A, B and C as follows:

$$\begin{aligned} A &= \{\text{the first die shows an even number}\}, \\ B &= \{\text{the second die shows an odd number}\}, \\ C &= \{\text{the two dice show both odd or both even number}\}. \end{aligned}$$

We have $P(A) = P(B) = P(C) = 1/2$ and $P(A \cap B) = P(B \cap C) = P(A \cap C) = 1/4$. Hence the three events are all pairwise independent. However, since $P(A \cap B \cap C) = 0 \neq 1/8 = P(A)P(B)P(C)$, the three events A, B and C are not independent.

Example 1.4.12. Assume that we toss two dice again. Define the event A, B and C as follows:

$$
\begin{aligned}
A &= \{\text{the second die shows 1,2 and 6}\},\\
B &= \{\text{the second die shows 4,5 and 6}\},\\
C &= \{\text{the sum of two dice equals 9}\} = \{(3,6),(4,5),(5,4),(6,3)\}.
\end{aligned}
$$

Then we have

$$
\begin{aligned}
A \cap B &= \{(i,j) | j = 6\}\\
A \cap C &= \{(3,6)\}\\
B \cap C &= \{(3,6),(4,5),(5,4)\}\\
A \cap B \cap C &= \{(3,6)\}
\end{aligned}
$$

and so it holds

$$P(A \cap B \cap C) = \frac{1}{36} = \frac{1}{2} \times \frac{1}{2} \times \frac{1}{9} = P(A)P(B)P(C).$$

But, since

$$P(A \cap B) = \frac{1}{6} \neq \frac{1}{2} \times \frac{1}{2} = P(A)P(B),$$

three event A, B and C are not mutually independent.

Theorem 1.4.13. *(i)* $P(B^c|A) = 1 - P(B|A)$.
(ii) If A and B are independent, so are A and B^c.

PROOF. (i) We have

$$\begin{aligned}
1 = P(S|A) &= P(B \cup B^c|A) \\
&= \frac{P((B \cup B^c) \cap A)}{P(A)} \\
&= \frac{P((B \cap A) \cup (B^c \cap A))}{P(A)} \\
&= \frac{P(B \cap A) + P(B^c \cap A)}{P(A)} \\
&= \frac{P(B \cap A)}{P(A)} + \frac{P(B^c \cap A)}{P(A)} \\
&= P(B|A) + P(B^c|A).
\end{aligned}$$

Alternatively, noting that $P(\cdot|A)$ is a probability measure, we can see that (i) is true by Theorem 1.2.3.

(ii) If A and B are independent, then $P(B|A) = P(B)$. Using (i), we have

$$\begin{aligned}
P(B^c|A) &= 1 - P(B|A) \\
&= 1 - P(B) \\
&= P(B^c).
\end{aligned}$$

This completes the proof.

□

Example 1.4.14. Let A and B are two independent events in a sample space and the probability that A occurs but B does not occur is 0.24 while the probability that neither occur is 0.36. Find $P(A)$ and $P(B)$.

SOLUTION. By Theorem 1.4.13, A and B^c are also independent. Thus from the problem statement, we have

$$0.24 = P(A \cap B^c) = P(A)P(B^c) = P(A)(1 - P(B)),$$

and so

$$P(A) - P(A)P(B) = 0.24. \tag{1.4.3}$$

From again the problem statement, we have

$$0.36 = P((A \cup B)^c) = 1 - P(A \cup B) = 1 - P(A) - P(B) + P(A \cap B),$$

and so

$$1 - P(A) - P(B) + P(A)P(B) = 0.36. \tag{1.4.4}$$

Adding (1.4.3) and (1.4.4), we obtain $1 - P(B) = 0.60$ and hence $P(B) = 0.40$ and $P(A) = 0.40$.

Alternatively, since

$$P(B) = 1 - P((A \cup B)^c) - P(A \cap B^c) = 1 - 0.36 - 0.24 = 0.4,$$

we have

$$0.24 = P(A \cap B^c) = P(A)P(B^c) - P(A)(0.6)$$

and hence $P(A) = 0.4$.

CHAPTER 2

Random Variables and Their Probability Distributions

2.1. Random variables and distributions

In describing the sample space of an experiment we did not specify that an individual outcome needs to be a number. However, we are usually interested not in the outcome itself, but rather in some quantity that depends upon the outcome. For example, in n tosses of a coin, we may be interested in the total number of heads and not in the specific order in which heads and tails occur. For this purpose we introduce the notion of a function defined on a sample space.

Definition 2.1.1. A **random variable** X is a function that assigns a real number $X(s)$ to each outcome s in the sample space S. More precisely, a random variable X is a function defined on S such that for every number x there exists the probability $P(\{s \in S| \ X(s) \leq x\})$.

In this definition x moves in the range space $\mathcal{R}(X)$ of values that the random variable X can take. Often a random variable is simply equal to the sample point - for example, the sample space representing the number of spots that turn up when a pair die is tossed is $S = \{1, 2, 3, 4, 5, 6\}$, and the random variable X which is the number of spots that turn up is given by $X(s) = s$ for all $s \in S$. For simplicity, we shall denote $P(\{s \in S |\ X(s) \leq x\})$ by $P(X \leq x)$.

Definition 2.1.2. The **distribution** or **law** of a random variable X is a set function P_X defined by

$$P_X(A) = P(X \in A) \quad \text{for any} \quad A \subset \mathcal{R}(X).$$

The **(cumulative) distribution function** of X is a function F_X defined by

$$F_X(x) = P_X((-\infty, x]) = P(X \leq x) \quad \text{for} \quad -\infty < x < \infty.$$

The function $S_X(x) = 1 - F_X(x) = P(X > x)$ is called the **survival function** or **reliability function** of X.

The axioms of probability and their properties imply that the distribution function has the following properties:

(i) $0 \leq F_X(x) \leq 1$.

(ii) $\lim_{x \to \infty} F_X(x) = 1, \quad \lim_{x \to -\infty} F_X(x) = 0$.

(iii) $F_X(x)$ is a nondecreasing function of x.

(iv) $F_X(x)$ is right continuous.

(v) $P(a < X \leq b) = F_X(b) - F_X(a)$.

Let us prove the property (ii).

$$\begin{aligned}\lim_{x \to \infty} F_X(x) &= P(X < \infty) \\ &= P(\{s \in S |\ X(s) < \infty\}).\end{aligned}$$

Put $B = \{s \in S |\ X(s) < \infty\}$. Clearly $B \subset S$. If $s \in S$, then $X(s) < \infty$ because X is a random variable. Hence $S \subset B$ and so $B = S$. Therefore

$$\lim_{x \to \infty} F_X(x) = P(S) = 1.$$

To show the property (v) we note that since

$$\{X \leq a\} \cup \{a < X \leq b\} = \{X \leq b\}$$

and since the two events on the left hand side are mutually exclusive, we have

$$F_X(a) + P(a < X \leq b) = F_X(b)$$

by Axiom III and the definition of the distribution. The property (v) then follows.

Definition 2.1.3. Let X be a random variable. If the range space $\mathcal{R}(X)$ of X is finite or countably infinite, we call X a **discrete random variable**.

Let X be a discrete random variable. Then the range space $\mathcal{R}(X)$ consists of at most a countably infinite number of values, $x_1, x_2, \cdots$. A function p_X defined by

$$p_X(x_i) = P(X = x_i) \quad \text{for any} \quad x_i \in \mathcal{R}(X).$$

is called the **probability mass function**(pmf) of X. The pmf p_X will satisfy two simple properties:

(a) $0 \leq p_X(x_i) \leq 1$ for all i

(b) $\sum_{i=1}^{\infty} p_X(x_i) = 1$.

Conversely, it is known that any function satisfying these two properties is a pmf of some discrete random variable. It is easy to show that

$$F_X(x) = \sum_{x_i \leq x} p_X(x_i) \quad \text{for any } x.$$

Example 2.1.4. One fair die is tossed repeatedly until the first even number turns up. Let X be the number of the tosses until the first even number turns up. Find the probability that X is an odd number.

SOLUTION. X is a discrete random variable that can take on a positive integer. Let A be the set of all odd numbers. We have to find $P(X \in A)$. If we denote by A_i the event that i-th throw is even, then $P(A_i) = P(A_i^c) =$

$1/2$ for all $i = 1, 2, \cdots$ and so pmf $p_X(n) = P(X = n)$ is given by

$$\begin{aligned} p_X(n) &= P(A_1^c \cap A_2^c \cap \cdots A_{n-1}^c \cap A_n) \\ &= P(A_1^c) \cdot P(A_2^c) \cdots P(A_{n-1}^c) \cdot P(A_n) \\ &= \left(\frac{1}{2}\right)^{n-1} \left(\frac{1}{2}\right) = \frac{1}{2^n}. \end{aligned}$$

Thus we have

$$\begin{aligned} P(X \in A) &= P(\{X = 1\} \cup \{X = 3\} \cup \{X = 5\} \cup \cdots) \\ &= p_X(1) + p_X(3) + p_X(5) + \cdots \\ &= \frac{1}{2} + \left(\frac{1}{2}\right)^3 + \left(\frac{1}{2}\right)^5 + \cdots \\ &= \frac{1}{2}\left\{1 + \left(\frac{1}{2}\right)^2 + \left(\frac{1}{2}\right)^4 + \cdots\right\} \\ &= \frac{1}{2}\left\{\frac{1}{1-(1/2)^2}\right\} = \frac{2}{3}. \end{aligned}$$

Definition 2.1.5. X is called a **continuous random variable** if $F_X(x)$ is absolute continuous, that is, if there exists a nonnegative function $f_X(x)$ such that for every real number x

$$F_X(x) = \int_{-\infty}^{x} f_X(u)du.$$

Here $f_X(x)$ is called the **probability density function**(pdf) of X.

Note that the pdf $f_X(x)$ of X satisfies the following two properties:

(a) $f_X(x) \geq 0$ for all x,

(b) $\int_{-\infty}^{+\infty} f_X(x)dx = 1$

and, conversely, any function satisfying these two properties is the pdf of some continuous random variable. It is easy to show that for any $-\infty < a \leq b < +\infty$,

$$P(a \leq X \leq b) = \int_a^b f_X(x)dx. \tag{2.1.1}$$

Remark 2.1.6. Let X be a continuous random variable.
(1) From the equality (2.1.1), we have for any $-\infty < a < +\infty$,

$$P(X = a) = P(a \le X \le a) = \int_a^a f_X(x)dx = 0$$

and hence

$$P(a < X < b) = P(a \le X < b) = P(a < X \le b) = P(a \le X \le b).$$

(2) By Definition 2.1.5, we see that F_X is continuous, non-decreasing, differentiable function such that $\frac{d}{dx}F_X(x) = f_X(x)$.
(3) A function $h_X(x) = -\frac{d}{dx}\ln S_X(x) = -\frac{d}{dx}\ln\{1 - F_X(x)\} = f_X(x)/\{1 - F_X(x)\} = f_X(x)/S_X(x)$ is called the **hazard rate** or **failure rate** of X.

Example 2.1.7. Let X be the random variable representing the number of dots that appear when one fair die is rolled once. Then X is a discrete random variable with pmf $p_X(x) = P(X = x) = 1/6$ for $x = 1, 2, 3, 4, 5, 6$. So the distribution function is given by

$$F_X(x) = P(X \le x) = \sum_{x_k \le x} p_X(x_k) = \begin{cases} 0 & \text{if } x < 1 \\ \frac{1}{6} & \text{if } 1 \le x < 2 \\ \frac{2}{6} & \text{if } 2 \le x < 3 \\ \frac{3}{6} & \text{if } 3 \le x < 4 \\ \frac{4}{6} & \text{if } 4 \le x < 5 \\ \frac{5}{6} & \text{if } 5 \le x < 6 \\ 1 & \text{if } x \ge 6. \end{cases}$$

The graph of $F_X(x)$ shows that it jumps at points $x = 1, 2, \cdots, 6$ and each height is $1/6$.

Example 2.1.8. Suppose that X is a continuous random variable on the interval (0,1) with the pdf

$$f_X(x) = \begin{cases} 2x, & 0 < x < 1 \\ 0, & \text{elsewhere.} \end{cases}$$

(i) Computer $P(X \leq 1/2 | 1/3 \leq X \leq 2/3)$. (ii) Find the distribution function of X.

SOLUTION. (i) Directly applying the definition of conditional probability, we have

$$\begin{aligned} P\left(X \leq \frac{1}{2} \mid \frac{1}{3} \leq X \leq \frac{2}{3}\right) &= \frac{P(1/3 \leq X \leq 1/2)}{P(1/3 \leq X \leq 2/3)} \\ &= \frac{\int_{1/3}^{1/2} 2x dx}{\int_{1/3}^{2/3} 2x dx} = \frac{5/36}{1/3} \\ &= \frac{5}{12}. \end{aligned}$$

(ii) The distribution function is given by

$$F_X(x) = \begin{cases} 0 & \text{if} \quad x \leq 0, \\ \int_0^x 2t dt = x^2 & \text{if} \quad 0 < x \leq 1, \\ 1 & \text{if} \quad x > 1. \end{cases}$$

Example 2.1.9. Suppose that the pdf of a continuous random variable X is given by

$$f_X(x) = Ke^{-|x|} \qquad -\infty < x < \infty.$$

Find the constant K, and then find the probability $P(|X| < a)$.

SOLUTION. Using Definition 2.1.2(b) to find K, it holds

$$1 = \int_{-\infty}^{\infty} Ke^{-|x|} dx = 2\int_0^{\infty} Ke^{-x} dx = 2K.$$

and so K=1/2. Now we have

$$\begin{aligned} P(|X| < a) &= \frac{1}{2}\int_{-a}^{a} e^{-|x|}dx \\ &= 2\left(\frac{1}{2}\right)\int_{0}^{a} e^{-x}dx \\ &= 1 - e^{-a} \end{aligned}$$

Definition 2.1.10. A random variable of which the distribution function jumps at some discrete points and is continuous and strictly increasing on some intervals is said to have a **mixed distribution**. The sum of the probability masses at the discrete values of the random variable where the distribution function jumps and the integral of the density function on continuous region of values of the random variable where the distribution function is continuous and strictly increasing must be 1.

Example 2.1.11. Suppose that X has probability 0.5 and 0.4 at $X = 0$ and $X = 1$, respectively, and that X is a continuous random variable with density function $f_X(x) = x/5$ for $0 < x < 1$, and that X has no probability mass or density elsewhere. This random variable mixed distribution. In fact, the range space is the interval [0,1] and

$$P(X = 0) + P(X = 1) + \int_{0}^{1} \frac{x}{5}dx = 0.5 + 0.4 + 0.1 = 1.$$

In this case we have $P(0 < X < 0.5) = \int_0^{0.5}(x/5)dx = 0.05$, but $F_X(0.5) = P(0 \leq X < 0.5) = P(X = 0) + \int_0^{0.5}(x/5)dx = 0.5 + 0.05 = 0.55$.

2.2. Expectation and variance

When a large collection of numbers is assembled, as in census, we are usually interested not in the individual numbers, but rather in certain descriptive quantities such as the average or the median. In this section we shall study two such descriptive quantities: the *expectation* and the *variance*.

Definition 2.2.1. Let X be a discrete random variable with possible values $x_1, x_2, \cdots$. Then the **expectation** $E[X]$ of X is defined as

$$E[X] = \sum_{i=1}^{\infty} x_i p_X(x_i) \tag{2.2.1}$$

if the series $\sum_{i=1}^{\infty} x_i p_X(x_i)$ converges absolutely, i.e., if $\sum_{i=1}^{\infty} |x_i| p_X(x_i) < \infty$. The expectation of X is also called the **expected value** or the **mean value** of X.

If the range space $\mathcal{R}(X)$ of a discrete random variable X is finite, the equality (2.2.1) becomes $E[X] = \sum_{i=1}^{n} x_i p_X(x_i)$. This may be considered as a "weighted average" of the possible values $x_1, x_2, \cdots, x_n$. If all these possible values are equally probable, then it holds $E[X] = (1/n)\sum_{i=1}^{n} x_i$. For example, if a fair die is tossed and the random variable X is the number of points showing, then $E[X] = \frac{1}{6}(1+2+3+4+5+6) = \frac{7}{2}$.

Example 2.2.2. Two students a and b, in order, toss an ordinary die. The student who toss the first number "3" wins. The game continuous until someone tosses number "3". Let X be the number of tosses by two students until the student b wins. Find the expectation $E[X]$ of X.

SOLUTION. If we denote by A_i the event that i-th throw is the number "3", then $P(A_i) = 1/6$ and $P(A_i^c) = 5/6$ for all $i = 1, 2, \cdots$ and so pmf

$p_X(n) = P(X = n)$ is given by

$$\begin{aligned} p_X(n) &= P(A_1^c \cap A_2^c \cap \cdots A_{n-1}^c \cap A_n) \\ &= P(A_1^c) \cdot P(A_2^c) \cdots P(A_{n-1}^c) \cdot P(A_n) \\ &= \left(\frac{5}{6}\right)^{n-1} \left(\frac{1}{6}\right). \end{aligned}$$

Since X can take values $2, 4, 6, \cdots$, by Definition 2.2.1, we have

$$\begin{aligned} E[X] &= \sum_{k=1}^{\infty} 2kP(X = 2k) \\ &= \sum_{k=1}^{\infty} 2kp_X(2k) \\ &= \sum_{k=1}^{\infty} 2k \left(\frac{5}{6}\right)^{2k-1} \left(\frac{1}{6}\right) \\ &= 2 \cdot \frac{5}{6} \cdot \frac{1}{6} + 4 \cdot \left(\frac{5}{6}\right)^3 \cdot \frac{1}{6} + 6 \cdot \left(\frac{5}{6}\right)^5 \cdot \frac{1}{6} + \cdots \\ &= \frac{1}{6} \left\{ 2 \cdot \frac{5}{6} + 4 \cdot \left(\frac{5}{6}\right)^3 + 6 \cdot \left(\frac{5}{6}\right)^5 + \cdots \right\} \end{aligned}$$

Here we use the fact

$$\begin{aligned} 2x + 4x^3 + 6x^5 + \cdots &= \frac{d}{dx}\left(x^2 + x^4 + x^6 + \cdots\right) \\ &= \frac{d}{dx}\left(1 + x^2 + x^4 + x^6 + \cdots\right) \\ &= \frac{d}{dx}\left\{\frac{1}{1 - x^2}\right\} \\ &= \frac{2x}{(1 - x^2)^2} \end{aligned}$$

Then we see that

$$E[X] = \frac{1}{6} \times \frac{2 \times (5/6)}{\{1 - (5/6)^2\}^2} = \frac{360}{121}$$

Definition 2.2.3. Let X be a continuous random variable with pdf f_X. The **expectation** $E[X]$ of X is defined as

$$E[X] = \int_{-\infty}^{\infty} x f_X(x)dx, \tag{2.2.2}$$

provided that $\int_{-\infty}^{\infty} |x| f_X(x)dx < \infty$.

Theorem 2.2.4. *Let X be a random variable and let h be a function.*
(i) If X is discrete, then we have

$$E[h(X)] = \sum_{i=1}^{\infty} h(x_i)p_X(x_i). \tag{2.2.3}$$

(ii) If X is a continuous random variable with pdf f_X, then we have

$$E[h(X)] = \int_{-\infty}^{\infty} h(x) f_X(x)dx. \tag{2.2.4}$$

PROOF. We shall prove (i) only. The proof of (ii) is somewhat more intricate. Let X be discrete, and suppose that $P(X \in A) = 1$. If $y = h(x)$ is one-to-one mapping of A onto some set B, then it holds for any $y \in B$

$$p_{h(X)}(y) = P(h(X) = y) = P(X = h^{-1}(y)) = P(X = x) = p_X(x).$$

Thus, by Definition 2.2.1, we have

$$E[h(X)] = \sum_{y \in B} y p_{h(X)}(y) = \sum_{x \in A} h(x)p_X(x).$$

□

Remark 2.2.5. Suppose that X has a mixed distribution with positive probability masses $P(X = a)$and $P(X = b)$ and a continuous pdf $f_X(x)$ on the interval (a, b) such that

$$P(X = a) + P(X = b) + \int_a^b f_X(x)dx = 1.$$

Then we define the expectation of X by

$$E[X] = aP(X = a) + bP(X = b) + \int_a^b x f_X(x)dx.$$

Definition 2.2.6. The function $h(x) = x^n$, where n is a positive integer, is of special importance. If $E[X^n]$ exists, we call it the **n-th moment** of X. If $E[X] = \mu_X$, then $E[(X - \mu_X)^n]$ is called the **n-th central moment of X about the mean μ_X**. In the case $n = 2$, the second central moment of X about the mean μ_X is called the **variance** of X and we denote it by $Var(X)$ or σ_X^2, i.e.,

$$Var(X) = E[(X - E[X])^2] = E[X^2] - E[X]^2.$$

The variance is a measure of the "dispersion" of X about the mean value. The **standard deviation** of X is the square root of the variance of X, and is denoted $\sigma_X = \sqrt{Var(X)}$.

Remark 2.2.7. (1) Let A be an event and

$$X(s) = I_A(s) = \begin{cases} 1 & \text{if} \quad s \in A, \\ 0 & \text{if} \quad s \in A^c. \end{cases}$$

Then $E[X] = P(A)$. In fact,

$$\begin{aligned} E[X] &= E[I_A] \\ &= 1 \cdot P(I_A = 1) + 0 \cdot P(I_A = 0) \\ &= P(I_A = 1) \\ &= P(\{s \in S | I_A(s) = 1\}) \\ &= P(\{s \in S | s \in A\}) \\ &= P(A). \end{aligned}$$

Here $I_A(s)$ is called the **indicate function** of A.
(2) If $0 < \alpha < 1$, then the **100α-th percentile** of the distribution of a random variable X is the number x_α which satisfies both of the following inequalities:

$$P(X \leq x_\alpha) \geq \alpha \quad \text{and} \quad P(X \geq x_\alpha) \geq 1 - \alpha.$$

For the case of a continuous random variable, x_α is the number satisfying only $P(X \leq x_\alpha) = \alpha$. If $\alpha = 0.5$, the 50th percentile of a distribution is referred to as the **median** of the distribution and is denoted by M, i.e., $P(X \leq M) = 0.5$.

Any point at that pmf or pdf of X is maximized is called the **mode** of X.

If X has the mean μ_X and the variance σ_X^2, then the **skewness** is defined as $E\left[(X - \mu_X)^3\right] / \sigma_X^3$.

When X is a continuous random variable with pdf f_X, we say that X is **symmetric** about a point a if $f_X(a + x) = f_X(a - x)$ for any value x. In this case, it follows that the expectation of X is 0 and the median of X is a.

(3) For any constants a and b and random variable X, we see easily $E[aX + b] = aE[X] + b$ and $Var(aX + b) = a^2 Var(X)$. These equalities imply that if $E[X] = \mu_X$ and $Var(X) = \sigma_X^2$, then

$$E\left[\frac{X - \mu_X}{\sigma_X}\right] = 0 \quad \text{and} \quad Var\left(\frac{X - \mu_X}{\sigma_X}\right) = 1.$$

(4) The **moment generating function**(mgf) $M_X(t)$ of the random variable X is defined by $M_X(t) = E\left[e^{tX}\right]$, which is either $\sum_x e^{tx} p_X(x)$ or $\int_{-\infty}^{\infty} e^{tx} f_X(x)dx$ if X is discrete or continuous, respectively. We see easily that

(i) $M_X(0) = 1$,

(ii) $M'_X(0) = E[X]$,

(iii) $\frac{d^n M_X}{dt^n}(0) = E[X^n]$ for $n = 2, 3, \cdots$, and hance

(iv) $Var(X) = M''_X(0) - \{M'_X(0)\}^2$.

Theorem 2.2.8. *Let X be a continuous random variable with the pdf f_X defined on the interval $[a, b)$ ($f_X(x) = 0$ if $x \leq a$), where $-\infty < a$ and*

$b \leq \infty$, and let F_X be the distribution function of X. Then

$$E[X] = a + \int_a^b (1 - F_X(x))dx. \tag{2.2.5}$$

PROOF. We shall prove only the equality (2.2.5) in the case that $b = \infty$, i.e.,

$$E[X] = a + \int_a^\infty (1 - F_X(x))dx,$$

since the proof of finite case is similar and easier. Using the integration by parts, we obtain

$$\begin{aligned}
E[X] &= \int_a^\infty x f_X(x)dx \\
&= \lim_{u\to\infty} \int_a^u x f_X(x)dx \\
&= \lim_{u\to\infty} \int_a^u x F'_X(x)dx \\
&= \lim_{u\to\infty} \left\{ [xF_X(x)]_a^u - \int_a^u F_X(x)dx \right\} \\
&= \lim_{u\to\infty} \left\{ uF_X(u) - \int_a^u F_X(x)dx \right\} \quad \text{since} \quad F_X(a) = 0 \\
&= \lim_{u\to\infty} \left\{ a - u(1 - F_X(u) + u - a - \int_a^u F_X(x)dx \right\} \\
&= \lim_{u\to\infty} \left\{ a - u(1 - F_X(u) + \int_a^u 1dx - \int_a^u F_X(x)dx \right\} \\
&= \lim_{u\to\infty} \left\{ a - u(1 - F_X(u)) + \int_a^u (1 - F_X(x))dx \right\}.
\end{aligned}$$

But

$$\begin{aligned}
u(1 - F_X(u)) &= u(1 - P(X \leq u)) \\
&= uP(X \geq u) \\
&= u \int_u^\infty f_X(x)dx \\
&< \int_u^\infty x f_X(x)dx \\
&= \int_a^\infty x f_X(x)dx - \int_a^u x f_X(x)dx,
\end{aligned}$$

and, since $\int_a^\infty x f_X(x)dx = E[|X|] < \infty$, the right hand side converges to 0 as $u \to \infty$. So it follows that

$$\lim_{u\to\infty} u(1 - F_X(u)) = 0.$$

Thus we have

$$\begin{aligned} E[X] &= a + \lim_{u\to\infty} \int_a^u (1 - F_X(x))dx \\ &= a + \int_a^\infty (1 - F_X(x))dx. \end{aligned}$$

The proof is complete. □

Theorem 2.2.9 (Jensen's inequality). *If a real valued function $h(x)$ is convex, i.e., $h''(x) \geq 0$ for all x, then*

$$E[h(X)] \geq h(E[X])$$

provided that the expectations exist and are finite.

PROOF. By Taylor's series expansion for $h(x)$ about $\mu = E[X]$,

$$h(x) = h(\mu) + h'(\mu)(x - \mu) + \frac{h''(\xi)(x - \mu)^2}{2}$$

where ξ is some value between x and μ. Since $h''(x) \geq 0$, we obtain

$$h(x) \geq h(\mu) + h'(\mu)(x - \mu).$$

Thus

$$h(X) \geq h(\mu) + h'(\mu)(X - \mu).$$

Taking expectations yields

$$E[h(X)] \geq h(\mu) + h'(\mu)E[(X - \mu)] = h(\mu)$$

and this completes the proof. □

Example 2.2.10. Let $h(x) = x^2$. Then $h''(x) = 2 \geq 0$ for all x, so that $E[X^2] \geq E[X]^2$. This is true since $E[X^2] - E[X]^2 = Var(X) \geq 0$. As another examples of convex functions are $h(x) = e^{kx}$ for all x and $h(x) = -x^{1/k}$ for $x \geq 0$.

Theorem 2.2.11 (Chebyshev's inequality). *Let X be a random variable with $E[X] = \mu$ and let c be any real number. Then, if ϵ is any positive number and $E[(X-c)^2] < \infty$, we have*

$$P(|X-c| \geq \epsilon) \leq \frac{E[(X-c)^2]}{\epsilon^2}. \tag{2.2.6}$$

The following forms, equivalent to (2.2.6), are immediate:
(i) Choosing $c = E[X] = \mu$ we obtain

$$P(|X-\mu| \geq \epsilon) \leq \frac{Var(X)}{\epsilon^2}. \tag{2.2.7}$$

(ii) Choosing $c = E[X] = \mu$ and $\epsilon = k\sigma$, where $\sigma^2 = Var(X)$, we obtain

$$P(|X-\mu| \geq k\sigma) \leq \frac{1}{k^2}. \tag{2.2.8}$$

This last form (2.2.8) is particularly indicative of how the variance measures the "degree of concentration" of probability near μ.

PROOF. We shall deal only with the continuous case. In the discrete case the argument is very similar with integrals replaced by sums. Let $f_X(x)$ be the pdf of X. By the definition of the expectation of a continuous random variable, we have

$$\begin{aligned}
E[(X-c)^2] &= \int_{-\infty}^{\infty} (x-c)^2 f_X(x)dx \\
&= \int_{\{x:|x-c|\geq\epsilon\}} (x-c)^2 f_X(x)dx + \int_{\{x:|x-c|<\epsilon\}} (x-c)^2 f_X(x)dx \\
&\geq \int_{-\infty}^{\infty} \epsilon^2 I_{\{x:|x-c|\geq\epsilon\}}(x) f_X(x)dx \\
&= \epsilon^2 E[I_{(|X-c|\geq\epsilon)}] \\
&= \epsilon^2 P(|X-c| \geq \epsilon) \quad \text{by Remark 2.2.7(1).}
\end{aligned}$$

This completes the proof. □

Example 2.2.12. Taking $k = 3/2$ in Chebyshev's inequality (2.2.8), we obtain

$$P(|X - \mu| \geq \frac{3}{2}\sigma) \leq \frac{4}{9} = 0.44.$$

It is possible to give more precise statement than Chebyshev's inequality, at least in some cases, if we know higher-order moments. For example, assume that X is a continuous random variable with pdf $f_X(x) = \sqrt{3}/2$ if $1 - 1/\sqrt{3} < x < 1 + 1/\sqrt{3}$, and $= 0$ otherwise. Then $E[X] = 1$, $Var(X) = 1/9$ and thus

$$\begin{aligned} P(|X - \mu| \geq \frac{3}{2}\sigma) &= P(|X - 1| \geq \frac{1}{2}) \\ &= 1 - P(|X - 1| < \frac{1}{2}) \\ &= 1 - P(\frac{1}{2} < X < \frac{3}{2}) \\ &= 1 - \frac{\sqrt{3}}{2} = 0.134. \end{aligned}$$

However, in many problems no assumption concerning the specific distribution of the random variable is justified, and in such cases Chebyshev's inequality can give us important information about the behavior of the random variable.

Remark 2.2.13. For any random variable X, since $X \leq |X|$ and $-X \leq |X|$, we obtain $E[X] \leq E[|X|]$ and $-E[X] \leq E[|X|]$. Hence we have

$$|E[X]| \leq E[|X|].$$

Theorem 2.2.14 (Schwartz's inequality). *Assume that $E[X^2]$ and $E[Y^2]$ are finite. Then $E[XY]$ is finite and*

$$|E[XY]|^2 \leq E[|X|^2]E[|Y|^2].$$

PROOF. If X is 0, then the inequality is immediate. So assume that X is not 0. Then $E[|X|^2] > 0$. For any real number x, let

$$h(x) = E[(x|X| + |Y|)^2] = E[|X|^2]x^2 + 2E[|XY|]x + E[|Y|^2].$$

Since $h(x)$ is the expectation of a nonnegative random variable, it must be nonnegative for all x. The quadratic equation $h(x) = 0$ has either no real roots or one real repeated root. Thus the discriminant must be ≤ 0, that is,

$$E[|XY|]^2 - E[|X|^2]E[|Y|^2] \leq 0.$$

From this and Remark 2.2.13, the result follows. □

2.3. Some special discrete distributions

In this section we study some special discrete random variables which are characterized by frequently used pmf's. The results of this section will be of considerable use in theoretical as well as practical applications.

Definition 2.3.1. A discrete random variable X is said to have a **uniform distribution** on n real numbers $\{x_1, x_2, \cdots, x_n\}$ and write $X \sim U(x_1, x_2, \cdots, x_n)$ if its pmf is of the form

$$p_X(x_k) = P(X = x_k) = \frac{1}{n}, \quad k = 1, 2, \cdots, n.$$

Theorem 2.3.2. *If $X \sim U(x_1, x_2, \cdots, x_n)$, then*

$$E[X] = \frac{1}{n}\sum_{k=1}^{n} x_k, \quad Var(X) = \frac{1}{n}\sum_{k=1}^{n}(x_k - \overline{x})^2,$$

where $\overline{x} = \sum_{k=1}^{n} x_k/n$.

PROOF. By the definition of the discrete random variable, we obtain

$$E[X]=\sum_{k=1}^{n} x_k p_X(x_k)=\frac{1}{n}\sum_{k=1}^{n} x_k.$$

Similarly, we have

$$E[X^2]=\sum_{k=1}^{n} x_k^2 p_X(x_k)=\frac{1}{n}\sum_{k=1}^{n} x_k^2$$

and so

$$\begin{aligned}
Var(X)&=\frac{1}{n}\sum_{k=1}^{n} x_k^2-\left(\frac{1}{n}\sum_{k=1}^{n} x_k\right)^2\\
&=\frac{1}{n}\sum_{k=1}^{n} x_k^2-\overline{x}^2\\
&=\frac{1}{n}\sum_{k=1}^{n} x_k^2-2\overline{x}^2+\overline{x}^2\\
&=\frac{1}{n}\sum_{k=1}^{n} x_k^2-\frac{2}{n}\sum_{k=1}^{n} x_k\overline{x}+\frac{1}{n}\sum_{k=1}^{n}\overline{x}^2\\
&=\frac{1}{n}\sum_{k=1}^{n}\left(x_k^2-2x_k\overline{x}+\overline{x}^2\right)\\
&=\frac{1}{n}\sum_{k=1}^{n}(x_k-\overline{x})^2.
\end{aligned}$$

This completes the proof. □

Remark 2.3.3. If $x_k=k,\ k=1,2,\cdots,n$, in Theorem 2.3.2, we obtain

$$E[X]=\frac{n+1}{2},\quad E[X^2]=\frac{(n+1)(2n+1)}{6} \tag{2.3.1}$$

and

$$Var(X)=\frac{n^2-1}{12}. \tag{2.3.2}$$

We can see that these are proved from

$$\sum_{k=1}^{n} k=\frac{n(n+1)}{2}\quad\text{and}\quad\sum_{k=1}^{n} k^2=\frac{n(n+1)(2n+1)}{6}.$$

Definition 2.3.4. Let X be a discrete random variable with possible values $0, 1, 2, \cdots$. If the pmf is given by

$$p_X(k) = P(X = k) = \frac{e^{-\lambda}\lambda^k}{k!},$$

then we say that X has a **Poisson distribution** with parameter $\lambda > 0$ and write $X \sim P(\lambda)$.

Theorem 2.3.5. *If $X \sim P(\lambda)$, then*

$$E[X] = Var(X) = \lambda.$$

PROOF. By the definition of the expectation, we have

$$E[X] = \sum_{k=0}^{\infty} \frac{ke^{-\lambda}\lambda^k}{k!} = \sum_{k=1}^{\infty} \frac{e^{-\lambda}\lambda^k}{(k-1)!}.$$

Letting $i = k - 1$, we see that this becomes

$$E[X] = \sum_{i=0}^{\infty} \frac{e^{-\lambda}\lambda^{i+1}}{i!} = \lambda \sum_{i=0}^{\infty} \frac{e^{-\lambda}\lambda^i}{i!} = \lambda.$$

Similarly,

$$E[X^2] = \sum_{k=0}^{\infty} \frac{k^2e^{-\lambda}\lambda^k}{k!} = \sum_{k=1}^{\infty} k\frac{e^{-\lambda}\lambda^k}{(k-1)!}.$$

Again letting $i = k - 1$, we obtain

$$\begin{aligned} E[X^2] &= \sum_{i=0}^{\infty}(i+1)\frac{e^{-\lambda}\lambda^{i+1}}{i!} \\ &= \lambda\sum_{i=0}^{\infty} i\frac{e^{-\lambda}\lambda^i}{i!} + \lambda\sum_{i=0}^{\infty}\frac{e^{-\lambda}\lambda^i}{i!} \\ &= \lambda^2 + \lambda, \end{aligned}$$

since the first sum represents $E[X] = \lambda$ and the second sum equals one. Hence

$$Var(X) = E[X^2] - E[X]^2 = \lambda^2 + \lambda - \lambda^2 = \lambda.$$

Alternatively, since

$$\begin{aligned} M_X(t) &= \sum_{k=0}^{\infty} e^{kt}\frac{e^{-\lambda}\lambda^k}{k!} \\ &= \sum_{k=0}^{\infty} \frac{e^{-\lambda}(e^t\lambda)^k}{k!} \\ &= \sum_{k=0}^{\infty} \frac{e^{-e^t\lambda}(e^t\lambda)^k}{k!} e^{-\lambda}e^{e^t\lambda} \\ &= e^{e^t\lambda-\lambda} \\ &= e^{\lambda(e^t-1)}, \end{aligned}$$

we have $E[X] = M_X'(0) = \lambda$ and $E[X^2] = M_X''(0) = \lambda^2 + \lambda$, and so $Var(X) = \lambda$. The proof is complete. □

Example 2.3.6. The Poisson distribution is often used as a model for counting the number of events of a certain type that occur in a certain period of time. Suppose that X represents the number of customers arriving for service at bank in a one hour period and that X has a Poisson distribution. If the probability that no one arriving at bank in a one hour period is e^{-30}, then

$$e^{-30} = P(X=0) = \frac{e^{-\lambda}\lambda^0}{0!} = e^{-\lambda}$$

and so $\lambda = 30$. This show that $E[X] = 30$, i.e., the mean of the number of customers arriving for service at bank in a one hour period is 30.

An independent trial which can results in only two possible outcomes, success with probability p or failure with probability $1 - p = q$, is called a **Bernoulli trial**. If n Bernoulli trials are performed and X is the number of successes that occur, then X is a random variable taking an integer between 0 and n and its pmf is given by

$$p_X(k) = P(X=k) = \binom{n}{k} p^k(1-p)^{n-k}, \quad k = 0, 1, 2, \cdots, n \tag{2.3.3}$$

Definition 2.3.7. We say that X has a **binomial distribution** with parameters n and p and write $X \sim B(n,p)$ if its pmf is given by (2.3.3). In the case of $n=1$, i.e., $X \sim B(1,p)$, X is called the **Bernoulli random variable**.

Theorem 2.3.8. *If $X \sim B(n,p)$, then*

$$E[X] = np, \quad Var(X) = np(1-p).$$

PROOF. By the definition of the expectation of a discrete random variable, we have

$$\begin{aligned} E[X] &= \sum_{k=1}^{n} k p_X(k) \\ &= \sum_{k=1}^{n} k \binom{n}{k} p^k (1-p)^{n-k} \\ &= \sum_{k=1}^{n} k \frac{n}{k} \binom{n-1}{k-1} p p^{k-1} (1-p)^{(n-1)-(k-1)} \\ &= np \sum_{k=1}^{n} \binom{n-1}{k-1} p^{k-1} (1-p)^{(n-1)-(k-1)} \\ &= np. \end{aligned}$$

By the similar calculus, we have

$$\begin{aligned} E[X^2] &= E[X(X-1)+X] \\ &= E[X(X-1)] + E[X] \\ &= \sum_{k=2}^{n} k(k-1) p_X(k) + np \\ &= \sum_{k=2}^{n} k(k-1) \binom{n}{k} p^k (1-p)^{n-k} + np \end{aligned}$$

$$
\begin{aligned}
&= \sum_{k=2}^{n} k(k-1)\frac{n}{k}\binom{n-1}{k-1}p^k(1-p)^{n-k} + np \\
&= n\sum_{k=2}^{n}(k-1)\frac{n-1}{k-1}\binom{n-2}{k-2}p^2p^{k-2}(1-p)^{(n-2)-(k-2)} + np \\
&= n(n-1)p^2\sum_{k=2}^{n}\binom{n-2}{k-2}p^{k-2}(1-p)^{(n-2)-(k-2)} + np \\
&= n(n-1)p^2 + np
\end{aligned}
$$

and hence

$$
\begin{aligned}
Var(X) &= E[X^2] - E[X]^2 \\
&= n(n-1)p^2 + np - (np)^2 \\
&= np(1-p).
\end{aligned}
$$

Alternatively, since

$$
\begin{aligned}
M_X(t) &= \sum_{k=1}^{n} e^{tk}p_X(k) \\
&= \sum_{k=1}^{n} e^{tk}\binom{n}{k}p^k(1-p)^{n-k} \\
&= \sum_{k=1}^{n}\binom{n}{k}(pe^t)^k(1-p)^{n-k} \\
&= (pe^t + 1 - p)^n,
\end{aligned}
$$

we can see that $E[X] = M'_X(0) = np$ and $Var(X) = M''_X(0) - M'_X(0)^2 = np(1-p)$. The proof is complete. □

Suppose successive independent trials of an experiment are performed for which the probability of success in single particular trial is p. Let r be an integer and X the number of failures until the r-th success occurs. Then we have

$$
p_X(x) = P(X = x) = \binom{x+r-1}{x}p^r(1-p)^x \quad x = 0, 1, 2, \cdots. \tag{2.3.4}
$$

Definition 2.3.9. For fixed positive integer $r \geq 1$ and $0 < p < 1$, a random variable X with pmf given by (2.3.4) is said to have a **negative binomial distribution** with parameters r and p. We will use the notation write $X \sim NB(r,p)$.

Theorem 2.3.10. *If $X \sim NB(r,p)$, then*

$$E[X] = \frac{r(1-p)}{p}, \quad Var(X) = \frac{r(1-p)}{p^2}.$$

PROOF. To prove this theorem we need somewhat intricate properties for combinatorial principles, so we shall not prove here. □

In the special case when $r = 1$ in (2.3.4), we have

$$p_X(x) = P(X = x) = p(1-p)^x \quad x = 0, 1, 2, \cdots. \tag{2.3.5}$$

In this case, X represents the number of failures until the first success. A random variable X with pmf (2.3.5) is said to have a **geometric distribution** with parameter p and write $X \sim G(p)$. Clearly, if $X \sim G(p)$, then we have

$$E[X] = \frac{1-p}{p}, \quad Var(X) = \frac{1-p}{p^2}.$$

Theorem 2.3.11 (The lack of memory property). *The property $X \sim G(p)$ is equivalent to the following condition*

$$P(X = n + k|X \geq n) = P(X = k).$$

PROOF. By the definition of the conditional probability, we obtain

$$\begin{aligned} P(X = n + k|X \geq n) &= \frac{P(X = n + k, X \geq n)}{P(X \geq n)} \\ &= \frac{P(X = n + k)}{P(X \geq n)}. \end{aligned}$$

If $X \sim G(p)$, then we obtain

$$P(X = n + k) = p(1-p)^{n+k}$$

and

$$\begin{aligned}P(X \geq n) &= \sum_{j=n}^{\infty} P(X=j) \\ &= \sum_{j=n}^{\infty}(1-p)^j p \\ &= p(1-p)^n \sum_{j=0}^{\infty}(1-p)^j \\ &= p(1-p)^n \frac{1}{1-(1-p)} \\ &= (1-p)^n,\end{aligned}$$

and so

$$P(X=n+k|X \geq n) = \frac{p(1-p)^{n+k}}{(1-p)^n} = (1-p)^k p.$$

This completes the proof. □

In a group of N objects, M are of Type I and $N-M$ are of Type II. When n objects are randomly chosen without replacement from the group of N, let X be the number that are of Type I in the group of n. Then we have

$$p_X(x) = P(X=x) = \frac{\binom{M}{x}\binom{N-M}{n-x}}{\binom{N}{n}}. \tag{2.3.6}$$

Definition 2.3.12. A random variable X with pmf given by (2.3.6) is said to have a **hypergeometric distribution** with parameters $N > 0$, $0 < M < N$ and $1 < n < N$ and write $X \sim HG(N, M, n)$.

Theorem 2.3.13. *If $X \sim HG(N, M, n)$, then*

$$E[X] = \frac{nM}{N}, \quad Var(X) = \frac{nM(N-M)(N-n)}{N^2(N-1)}.$$

PROOF. This can be proved by the same method as the proof of Theorem 2.3.8. The details of the proof is left as an exercise. □

Example 2.3.14. Let X be the number of diamonds that a hand of 13 cards contained in a game of bridge. Find $E[X^2]$.

SOLUTION. Since $X \sim HG(52, 13, 13)$, by Theorem 2.3.13, we have

$$E[X] = \frac{nM}{N} = \frac{13 \times 13}{52} = \frac{13}{4}$$

and

$$\begin{aligned} Var(X) &= \frac{nM(N-M)(N-n)}{N^2(N-1)} \\ &= \frac{13 \times 13 \times 39 \times 39}{52^2 \times 51} \\ &= \frac{39 \times 13}{16 \times 17}. \end{aligned}$$

Hence we obtain

$$\begin{aligned} E[X^2] &= Var(X) + E[X]^2 \\ &= \frac{39 \times 13}{16 \times 17} + \frac{13^2}{4^2} = \frac{507 + 2873}{272} \\ &= 12.42. \end{aligned}$$

Finally we define the multinomial distribution. The binomial distribution is generalized in the following natural fashion. Suppose that an experiment is repeated n times. Each outcome of the experiment belongs to one of k mutually exclusive and exhaustive events $A_1, A_2, \cdots, A_k$. Let p_j for $j = 1, 2, \cdots, k$ be the probability that the event A_j occurs. We assume that all n outcomes are independent. Let n_j be the the number that event A_j occurs. Then the probability that event A_j occurs is given by

$$\frac{n!}{n_1!n_2!\cdots n_k!}p_1^{n_1}p_2^{n_2}\cdots p_k^{n_k}.$$

If $(X_1, X_2, \cdots, X_k)$ is a random vector such that $X_j = n_j$ means that event A_j has occurred n_j times, then the joint pmf of $(X_1, X_2, \cdots, X_k)$ is given by

$$P(X_1 = n_1, X_2 = n_2, \cdots, X_k = n_k)$$

$$= \begin{cases} \frac{n!}{n_1!n_2!\cdots n_k!} p_1^{n_1} p_2^{n_2} \cdots p_k^{n_k} & \text{if} \quad n = \sum_{j=1}^{k} n_j, \\ 0 & \text{otherwise.} \end{cases} \tag{2.3.7}$$

A random vector $(X_1, X_2, \cdots, X_k)$ with joint pmf given by (2.3.7) is said to have a **multinomial distribution** with parameters n and $p_1, p_2, \cdots, p_k$, and we denote

$$(X_1, X_2, \cdots, X_k) \sim M(n; p_1, p_2, \cdots, p_k)$$

2.4. Some special continuous distributions

In this section we study some special continuous random variables which are characterized by frequently used pdf's. Many of the random variables we shall now introduce have important applications.

Definition 2.4.1. A continuous random variable X is said to have a **uniform distribution** on the interval (a, b) and write $X \sim U(a, b)$ if its pdf is of the form

$$f_X(x) = \begin{cases} \frac{1}{b-a} & \text{if} \quad a < x < b, \\ 0 & \text{otherwise.} \end{cases}$$

Theorem 2.4.2. *If* $X \sim U(a, b)$, *then*

$$E[X] = \frac{a+b}{2}, \quad Var(X) = \frac{(b-a)^2}{12}.$$

PROOF. By the definition of the expectation of a continuous random variable, we have

$$\begin{aligned} E[X] &= \int_{-\infty}^{\infty} x f_X(x) dx \\ &= \int_a^b \frac{x}{b-a} dx \\ &= \frac{b^2 - a^2}{2(b-a)} \\ &= \frac{b+a}{2}. \end{aligned}$$

Now to find $Var(X)$, we first calculate $E[X^2]$.

$$\begin{aligned} E[X^2] &= \int_a^b \frac{x^2}{b-a} dx \\ &= \frac{b^3 - a^3}{3(b-a)} \\ &= \frac{b^2 + ab + a^2}{3}. \end{aligned}$$

Hence we have

$$\begin{aligned} Var(X) &= E[X^2] - E[X]^2 \\ &= \frac{b^2 + ab + a^2}{3} - \frac{(b+a)^2}{4} \\ &= \frac{(b-a)^2}{12}. \end{aligned}$$

The proof is complete. □

One of the most important continuous random variables is the following.

Definition 2.4.3. Let $-\infty < \mu < \infty$ and $\sigma > 0$. A continuous random variable X is said to have a **normal distribution** with parameters μ and σ^2 and write $X \sim N(\mu, \sigma^2)$ if its pdf is of the form

$$f_X(x) = \frac{1}{\sqrt{2\pi}\sigma} \exp\left\{ -\frac{1}{2} \left(\frac{x-\mu}{\sigma} \right)^2 \right\}, \quad -\infty < x < \infty. \tag{2.4.1}$$

In this case, X is also called a **normal random variable** with parameters μ and σ^2.

Theorem 2.4.4. *If $X \sim N(\mu, \sigma^2)$, then*

$$M_X(t) = \exp\left(\mu t + \frac{\sigma^2 t^2}{2}\right), \quad E[X] = \mu, \quad Var(X) = \sigma^2.$$

PROOF. By the definition of mgf, we have

$$\begin{aligned}
M_X(t) &= E[e^{tX}] \\
&= \int_{-\infty}^{\infty} e^{tx} f_X(x) dx \\
&= \int_{-\infty}^{\infty} e^{tx} \frac{1}{\sqrt{2\pi}\sigma} e^{\frac{-(x-\mu)^2}{2\sigma^2}} dx \\
&= \frac{1}{\sqrt{2\pi}\sigma} \int_{-\infty}^{\infty} \exp\left(\frac{-x^2 - \mu^2 + 2x\mu + 2\sigma^2 tx}{2\sigma^2}\right) dx \\
&= \frac{1}{\sqrt{2\pi}\sigma} \int_{-\infty}^{\infty} \exp\left(\frac{-x^2 + 2x(t\sigma^2 + \mu)}{2\sigma^2} - \frac{\mu^2}{2\sigma^2}\right) dx \\
&= \frac{1}{\sqrt{2\pi}\sigma} \int_{-\infty}^{\infty} \exp\left(\frac{-(x - \mu - \sigma^2 t)^2}{2\sigma^2} + \mu t + \frac{\sigma^2 t^2}{2}\right) dx \\
&= \exp\left(\mu t + \frac{\sigma^2 t^2}{2}\right) \frac{1}{\sqrt{2\pi}\sigma} \int_{-\infty}^{\infty} \exp\left(\frac{-(x - \mu - \sigma^2 t)^2}{2\sigma^2}\right) dx \\
&= \exp\left(\mu t + \frac{\sigma^2 t^2}{2}\right).
\end{aligned}$$

Next, by Remark 2.2.7, we have

$$E[X] = M'(t)|_{t=0} = (\mu + \sigma^2 t) \exp\left(\mu t + \frac{\sigma^2 t^2}{2}\right)\Big|_{t=0} = \mu$$

and

$$E[X^2] = M''(t)|_{t=0} = \sigma^2 + \mu^2.$$

Thus we have

$$\mathrm{Var}(X) = \sigma^2 + \mu^2 - \mu^2 = \sigma^2.$$

The proof is complete. □

Theorem 2.4.5. *Let $X \sim N(\mu, \sigma^2)$. If $Y = aX + b$, $-\infty < a, b < \infty$, $a \neq 0$, then $Y \sim N(a\mu + b, a^2\sigma^2)$.*

PROOF. Let F_X and F_Y be the distribution functions of X and Y, respectively. Then it is sufficient to show that

$$F_Y(y) = \int_{-\infty}^{y} \frac{1}{\sqrt{2\pi}|a|\sigma} \exp\left[-\frac{(t-(a\mu+b))^2}{2(a\sigma)^2}\right] dt. \tag{2.4.2}$$

First we assume that $a > 0$. Then

$$\begin{aligned} F_Y(y) &= P(Y \leq y) \\ &= P(aX + b \leq y) \\ &= P(X \leq \frac{y-b}{a}) \\ &= \int_{-\infty}^{\frac{y-b}{a}} \frac{1}{\sqrt{2\pi}\sigma} \exp\left[-\frac{(x-\mu)^2}{2\sigma^2}\right] dx. \end{aligned}$$

Letting $t = ax + b$, it holds

$$\begin{aligned} F_Y(y) &= \int_{-\infty}^{y} \frac{1}{\sqrt{2\pi}\sigma} \exp\left[-\frac{1}{2\sigma^2}\left(\frac{t-b}{a} - \mu\right)^2 \frac{1}{a}\right] dt \\ &= \int_{-\infty}^{y} \frac{1}{\sqrt{2\pi}a\sigma} \exp\left[-\frac{(t-(a\mu+b))^2}{2(a\sigma)^2}\right] dt \\ &= \int_{-\infty}^{y} \frac{1}{\sqrt{2\pi}|a|\sigma} \exp\left[-\frac{(t-(a\mu+b))^2}{2(a\sigma)^2}\right] dt \end{aligned}$$

Thus the equality (2.4.2) holds. Next, we assume $a < 0$. Then

$$\begin{aligned} F_Y(y) &= P(Y \leq y) \\ &= P(aX + b \leq y) \\ &= P(X \geq \frac{y-b}{a}) \ (a < 0) \\ &= 1 - P(X \leq \frac{y-b}{a}) \\ &= 1 - \int_{-\infty}^{\frac{y-b}{a}} \frac{1}{\sqrt{2\pi}\sigma} \exp\left[-\frac{(x-\mu)^2}{2\sigma^2}\right] dx. \end{aligned}$$

Letting $t = ax + b$, it holds

$$\begin{aligned} F_Y(y) &= 1 - \int_{\infty}^{y} \frac{1}{\sqrt{2\pi} a\sigma} \exp\left[-\frac{(t-(a\mu+b))^2}{2(a\sigma)^2}\right] dt \\ &= 1 - \int_{y}^{\infty} \frac{1}{\sqrt{2\pi}(-a)\sigma} \exp\left[-\frac{(t-(a\mu+b))^2}{2(a\sigma)^2}\right] dt \\ &= \int_{-\infty}^{y} \frac{1}{\sqrt{2\pi}(-a)\sigma} \exp\left[-\frac{(t-(a\mu+b))^2}{2(a\sigma)^2}\right] dt \\ &= \int_{-\infty}^{y} \frac{1}{\sqrt{2\pi}|a|\sigma} \exp\left[-\frac{(t-(a\mu+b))^2}{2(a\sigma)^2}\right] dt. \end{aligned}$$

That is, the equality (2.4.2) holds. The proof is complete. □

Remark 2.4.6. Taking $a = 1/\sigma$ and $b = -\mu/\sigma$ in Theorem 2.4.5, we get new random variable

$$\frac{X-\mu}{\sigma} \sim N(0, 1^2).$$

We call it a **standardized random variable** or **normalized version** of X.

Remark 2.4.7. If $Z \sim N(0, 1^2)$ we say that Z has a **standard normal distribution**. That is, the distribution function of Z may be written as

$$F_Z(z) = P(Z \leq z) = \int_{-\infty}^{z} \frac{1}{\sqrt{2\pi}} e^{-x^2/2} dx. \tag{2.4.3}$$

$F_Z(z)$ is sometimes denoted by $\Phi(z)$. Tables of values of $\Phi(z)$ may be found in most probability or statistics textbooks and are provided with the exam. Since the pdf $f_Z(z)$ of Z is symmetric with respect to 0, we see that

$$\Phi(-z) = 1 - \Phi(z)$$

If $0 < \alpha < 1$, the notation z_α denotes the point such that $P(Z > z_\alpha) = \alpha$ or $\Phi(z_\alpha) = P(Z \leq z_\alpha) = 1 - \alpha$. z_α is called the $100(1-\alpha)$**-th percentile** of the standard normal distribution.

Example 2.4.8. If $X \sim N(5, 4^2)$, find $P(0 \le X \le 10)$.

SOLUTION. Since $\frac{X-5}{4} \sim N(0, 1^2)$,

$$\begin{aligned} P(0 \le X \le 10) &= P\left(\frac{0-5}{4} \le \frac{X-5}{4} \le \frac{10-5}{4}\right) \\ &= P(-1.25 \le Z \le 1.25) \\ &= P(Z \le 1.25) - P(Z \le -1.25) \\ &= \Phi(1.25) - \Phi(-1.25) \\ &= \Phi(1.25) - [1 - \Phi(1.25)] \\ &= 2\Phi(1.25) - 1 \end{aligned}$$

From the standard normal table we have $\Phi(1.25) = 0.8944$ and so $P(0 \le X \le 10) = 2 \times 0.8944 - 1 = 0.7888$.

Example 2.4.9. If $X \sim N(60, 6^2)$, find x satisfying $P(X \le x) = 0.05$.

SOLUTION. Since

$$0.05 = P(X \le x) = P\left(\frac{X-60}{6} \le \frac{x-60}{6}\right)$$

and $\frac{X-60}{6} \sim N(0, 1^2)$, from the standard normal table, we have

$$\frac{x-60}{6} = -1.645.$$

Thus $x = 50.13$.

Remark 2.4.10. Let $X_i \sim N(\mu_i, \sigma_i^2)$, $i = 1, 2$. If X_1, X_2 are independent and $Y = a_1X_1 + a_2X_2$, $a_i \ne 0$, $i = 1, 2$, then by the same way of the proof of Theorem 2.4.5, we see that

$$Y \sim N(a_1\mu_1 + a_2\mu_2, a_1^2\sigma_1^2 + a_2^2\sigma_2^2).$$

Example 2.4.11. Let random variables $X_1, X_2, \cdots$ are independent and have normal distribution with mean 3 and variance 1, and let

$$S_n = X_1 + X_2 + \cdots + X_n.$$

Find the smallest number n such that $P(S_n > 60) \geq 0.8413$.

SOLUTION. Since $X_i \sim N(3, 1^2)$ for all $i = 1, 2, \cdots$, and these are independent, we know that

$$E[S_n] = \sum_{i=1}^{n} E[X_i] = 3n.$$

and

$$Var(S) = \sum_{i=1}^{n} Var(X_i) = n.$$

By Remark 2.4.10, $S_n \sim N(3n, n)$ and hence

$$\begin{aligned} 0.8413 &\leq P(S_n > 60) \\ &= P\left(\frac{S_n - 3n}{\sqrt{n}} > \frac{60 - 3n}{\sqrt{n}}\right) \\ &= 1 - P\left(\frac{S_n - 3n}{\sqrt{n}} \leq \frac{60 - 3n}{\sqrt{n}}\right) \\ &\approx 1 - \Phi\left(\frac{60 - 3n}{\sqrt{n}}\right) \end{aligned}$$

On the other hand, from the standard normal table, we see that

$$0.8413 = \Phi(1) = 1 - \Phi(-1).$$

Thus

$$1 - \Phi(-1) \leq 1 - \Phi\left(\frac{60 - 3n}{\sqrt{n}}\right),$$

that is,

$$\Phi\left(\frac{60 - 3n}{\sqrt{n}}\right) \leq \Phi(-1)$$

and hence

$$\frac{60 - 3n}{\sqrt{n}} \leq -1.$$

Rewriting this,

$$3n - \sqrt{n} - 60 \geq 0$$

or

$$(3\sqrt{n} + 14)(\sqrt{n} - 5) \geq 0.$$

From this we have $\sqrt{n} = 5$ and so $n = 25$.

Remark 2.4.12. If X is an integer valued random variable with mean μ and variance σ^2, then the probability $P(m \le X \le n)$ is approximated by assuming that X has a normal distribution and finding the probability $P(m - 0.5 \le X \le n + 0.5)$, i.e.,

$$\begin{aligned} P(m \le X \le n) &= P(m - 0.5 \le X \le n + 0.5) \\ &= P\left(\frac{m - 0.5 - \mu}{\sigma} \le \frac{X - \mu}{\sigma} \le \frac{n + 0.5 - \mu}{\sigma}\right) \\ &\approx \Phi\left(\frac{n + 0.5 - \mu}{\sigma}\right) - \Phi\left(\frac{m - 0.5 - \mu}{\sigma}\right) \end{aligned}$$

This procedure is called the **continuous correction**.

Example 2.4.13. If $X \sim B(40, 0.5)$, find $P(X = 20)$ by the continuous correction and compare it to the exact solution.

SOLUTION. By Theorem 2.3.8, $E[X] = 40 \times 0.5 = 20$ and $Var(X) = 40 \times 0.5 \times 0.5 = 10$. So we have

$$\begin{aligned} P(X = 20) &= P(19.5 \le X \le 20.5) \\ &= P\left(\frac{19.5 - 20}{\sqrt{10}} \le \frac{X - 20}{\sqrt{10}} \le \frac{20.5 - 20}{\sqrt{10}}\right) \\ &= P\left(-0.16 \le \frac{X - 20}{\sqrt{10}} \le 0.16\right) \\ &\approx \Phi(0.16) - \Phi(-0.16) \\ &= 2\Phi(0.16) - 1 \\ &= 0.1272. \end{aligned}$$

The exact value is

$$P(X = 20) = \binom{40}{20} 0.5^{20} 0.5^{40-20} \approx 0.1254.$$

Example 2.4.14. Let X be a normal random variable satisfying $P(X < 3) = 0.5$ and $P(X < 5) = 0.841$. Find the standard deviation σ of X.

SOLUTION. Since the pdf of the normal random variable symmetric about its mean μ, we have $\mu = 3$. Since

$$0.841 = P(X < 5) = P(\frac{X-3}{\sigma} < \frac{2}{\sigma})$$

and $\frac{X-3}{\sigma} \sim N(0, 1^2)$, it follows from the table for standard normal distribution that $\frac{2}{\sigma} = 1$ and so $\sigma = 2$.

Definition 2.4.15. A continuous random variable X is said to have a **log-normal distribution** with parameters μ and σ^2 and write $X \sim LN(\mu, \sigma^2)$ if its pdf is of the form

$$f_X(x) = \frac{1}{\sqrt{2\pi}\sigma x} \exp\left\{-\frac{1}{2}\left(\frac{\ln x - \mu}{\sigma}\right)^2\right\}, \quad 0 \leq x < \infty. \tag{2.4.4}$$

It should be noted that the parameters μ and σ^2 are not the mean and variance, respectively, of the random variable X. The logarithm of the random variable X is normally distributed with mean μ and variance σ^2. The mean and variance of the log-normal random variable are given as follows:

$$\begin{aligned} E[X] &= \exp\left\{\mu + \frac{\sigma^2}{2}\right\} \\ Var(X) &= \exp\{2\mu + \sigma^2\}(\exp\{\sigma^2\} - 1). \end{aligned} \tag{2.4.5}$$

Theorem 2.4.16. *Let $X \sim LN(\mu, \sigma^2)$ and $Y = e^a X^b$, where a and b are constants. Then $Y \sim LN(a + b\mu, b^2\sigma^2)$.*

PROOF. By the definition of Y, we can write $\ln Y = a + b \ln X$. Since X has the log-normal distribution with parameters μ and σ^2, $\ln X$ obeys the normal distribution with parameters μ and σ^2, i.e., $\ln X \sim N(\mu, \sigma^2)$.By applying Theorem 2.4.5, we have $\ln Y = a + b \ln X \sim N(a + b\mu, b^2\sigma^2)$. This shows that $Y \sim LN(a + b\mu, b^2\sigma^2)$. □

In the case that $a = 0$ and $b = -1$ in Theorem 2.4.16, we can see that $1/X \sim LN(-\mu, \sigma^2)$.

Definition 2.4.17. A continuous random variable X is said to have an **exponential distribution** with parameters $\lambda > 0$ and write $X \sim E(\lambda)$ if its pdf is of the form

$$f_X(x) = \begin{cases} \lambda e^{-\lambda x}, & \text{if } \quad x > 0 \\ 0, & \text{otherwise.} \end{cases} \tag{2.4.6}$$

It is easy to see that, for $X \sim E(\lambda)$, $F_X(x) = P(X \leq x) = 1 - e^{-\lambda x}$ for $x \geq 0$ and $M_X(t) = \frac{\lambda}{\lambda - t}$ for $t < \lambda$. So $E[X] = \frac{1}{\lambda}$ and $Var(X) = \frac{1}{\lambda^2}$. Thus, if $X \sim E(\lambda)$, then we also say that X has an exponential distribution with mean $1/\lambda$.

Example 2.4.18. Let $X \sim E(1/1,000)$ and $X_1, X_2, \cdots, X_{100}$ be a sequence of independent random variables having the same distribution with X. Define

$$M = \max(X_1, X_2, \cdots, X_{100}),$$
$$K = \min(X_1, X_2, \cdots, X_{100}).$$

Then find $P(M > 7200)$ and $P(K < 10)$.

SOLUTION. Since $X \sim E(1/1,000)$, the pdf is given by

$$f_X x) = \frac{1}{1000} e^{-\frac{1}{1000}x}$$

Thus we have

$$\begin{aligned}
P(M > 7200) &= 1 - P(M \le 7200) \\
&= 1 - P(X_1 \le 7200, X_2 \le 7200, \cdots, X_{100} \le 7200) \\
&= 1 - \Pi_{i=1}^{100} P(X_i \le 7200) \\
&= 1 - [F_X(7200)]^{100} \\
&= 1 - \left[\int_0^{7200} f_X(x)dx\right]^{100} \\
&= 1 - \left[\int_0^{7200} \frac{1}{1000} e^{-\frac{1}{1000}x} dx\right]^{100} \\
&= 0.071
\end{aligned}$$

and

$$\begin{aligned}
P(K < 10) &= 1 - P(K \ge 10) \\
&= 1 - P(X_1 \ge 10, X_2 \ge 10, \cdots, T_{100} \ge 10) \\
&= 1 - \Pi_{i=1}^{100} P(X_i \ge 10) \\
&= 1 - [P(X \ge 10)]^{100} \\
&= 1 - [1 - P(X \le 10)]^{100} \\
&= 1 - \left[1 - \int_0^{10} \frac{1}{1000} e^{-\frac{1}{1000}x} dx\right]^{100} \\
&= 0.63.
\end{aligned}$$

Theorem 2.4.19 (The lack of memory property). *The property $X \sim E(\lambda)$ is equivalent to the following condition*

$$P(X > x + y | X > x) = P(X > y)$$

for $x, y > 0$.

PROOF. By the definition of the conditional probability, we obtain

$$\begin{aligned}
P(X > x + y | X > x) &= \frac{P(X > x + y, X > x)}{P(X > x)} \\
&= \frac{P(X > x + y)}{P(X > x)}.
\end{aligned}$$

If $X \sim E(\lambda)$, then

$$P(X > x + y) = 1 - F_X(x + y) = e^{-\lambda(x+y)}$$

and

$$P(X > x) = e^{-\lambda x},$$

so that

$$P(X > x + y | X > x) = \frac{e^{-\lambda(x+y)}}{e^{-\lambda x}} = e^{-\lambda y}.$$

This completes the proof. □

Example 2.4.20. Let $Y_1, Y_2, \cdots, Y_n$ are independent random variable such that $Y_k \sim E(\lambda_k)$ for all $k = 1, 2, \cdots, n$. Verify that

$$\min\{Y_1, Y_2, \cdots, Y_n\} \sim E(\lambda_1 + \lambda_2 + \cdots + \lambda_n).$$

SOLUTION. Let $Y = \min\{Y_1, Y_2, \cdots, Y_n\}$. Then we have

$$\begin{aligned} F_Y(y) &= 1 - P(Y > y) \\ &= 1 - P(Y_1 > y, Y_2 > y, \cdots, Y_n > y) \\ &= 1 - P(Y_1 > y)P(Y_2 > y) \cdots P(Y_n > y) \\ &= 1 - e^{-\lambda_1 x} e^{-\lambda_2 x} \cdots e^{-\lambda_n x} = 1 - e^{-(\lambda_1 + \lambda_2 + \cdots + \lambda_n) y}. \end{aligned}$$

Thus

$$f_Y(y) = F_Y'(y) = (\lambda_1 + \lambda_2 + \cdots + \lambda_n) e^{-(\lambda_1 + \lambda_2 + \cdots + \lambda_n) y}.$$

This shows that $Y \sim E(\lambda_1 + \lambda_2 + \cdots + \lambda_n)$.

Definition 2.4.21. A continuous random variable X is said to have a **truncated exponential distribution** with parameters λ and L and write $X \sim TE(\lambda, L)$ if its distribution function is of the form

$$F_X(x) = \begin{cases} 0, & \text{if } x < 0 \\ 1 - e^{-\lambda x}, & \text{if } 0 \le x < L \\ 1, & \text{if } x \ge L. \end{cases} \tag{2.4.7}$$

Remark 2.4.22. When $X \sim TE(\lambda, L)$, the truncated exponential distribution function $F_X(x)$ jumps at a point $x = L$. The height is $e^{-\lambda L}$. Thus $F_X(x)$ is a mixed distribution with the pdf $f_X(x) = \lambda e^{-\lambda x}$ on an interval $(0, L)$, and pmf $p_X(L) = e^{-\lambda L}$ at a point $x = L$. Clearly it holds

$$\int_0^L f_X(x)dx + p_X(L) = \int_0^L \lambda e^{-\lambda x}dx + e^{-\lambda L} = -e^{-\lambda L} + 1 + e^{-\lambda L} = 1.$$

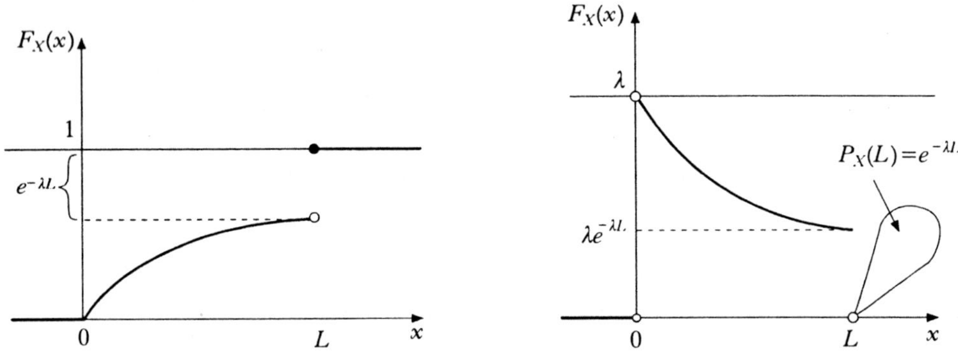

In this case, the expectation and variance of X are given by

$$E[X] = \int_0^L x\lambda e^{-\lambda x}dx + Le^{-\lambda L} = \frac{1 - e^{-\lambda L}}{\lambda}$$

and

$$\begin{aligned} Var(X) &= \int_0^L x^2\lambda e^{-\lambda x}dx + L^2e^{-\lambda L} - \left(\frac{1 - e^{-\lambda L}}{\lambda}\right)^2 \\ &= \frac{2}{\lambda^2}(1 - e^{-\lambda L}) - \frac{2L}{\lambda}e^{-\lambda L} - \left(\frac{1 - e^{-\lambda L}}{\lambda}\right)^2 \\ &= \frac{1 - 2\lambda Le^{-\lambda L} - e^{-2\lambda L}}{\lambda^2}. \end{aligned}$$

Definition 2.4.23. The **Gamma function**, denoted by Γ, is defined as follows:

$$\Gamma(\alpha) = \int_0^\infty x^{\alpha-1}e^{-x}dx, \quad \alpha > 0.$$

By the definition of $\Gamma(\alpha)$, we see that $\Gamma(\alpha) = (\alpha - 1)\Gamma(\alpha - 1)$. If α is a positive integer, then $\Gamma(\alpha) = (\alpha - 1)!$. And we can see that $\Gamma(\frac{1}{2}) = \sqrt{\pi}$. With the aid of the Gamma function we can introduce the Gamma distribution.

Definition 2.4.24. A continuous random variable X is said to have a **Gamma distribution** with parameters $\alpha > 0$ and $\beta > 0$, we write $X \sim G(\alpha, \beta)$, if its pdf is of the form

$$f_X(x) = \begin{cases} \frac{\beta^\alpha x^{\alpha-1} e^{-\beta x}}{\Gamma(\alpha)}, & \text{if} \quad x > 0 \\ 0, & \text{otherwise.} \end{cases} \tag{2.4.8}$$

Remark 2.4.25. (1) When $X \sim G(\alpha, \beta)$, the mgf of X is easily calculated by the following way.

$$\begin{aligned} M_X(t) &= E[e^{tX}] \\ &= \int_{-\infty}^{\infty} e^{tx} f_X(x) dx \\ &= \int_0^{\infty} \frac{\beta^\alpha x^{\alpha-1} e^{(t-\beta)x}}{\Gamma(\alpha)} dx, \quad t < \beta. \end{aligned}$$

Letting $y = (\beta - t)x$, the right hand side is equal to

$$\left(\frac{\beta}{\beta - t}\right)^\alpha \int_0^{\infty} \frac{y^{\alpha-1} e^{-y}}{\Gamma(\alpha)} dy = \left(\frac{\beta}{\beta - t}\right)^\alpha, \quad t < \beta.$$

Thus we obtain

$$E[X] = M'_X(0) = \frac{\alpha}{\beta}, \quad E[X^2] = M''_X(0) = \frac{\alpha(\alpha+1)}{\beta^2}$$

and hence

$$Var(X) = E[X^2] - E[X^2] = \frac{\alpha}{\beta^2}.$$

(2) If $\alpha = 1$, (2.4.8) becomes $f_X(x) = \beta e^{-\beta x}$. Hence the exponential distribution with parameter β is a special case of the Gamma distribution with parameter 1 and β.

(3) Another special case of the Gamma distribution is obtained if we let

$\beta = 1/2$ and $\alpha = n/2$, where n is positive integer. In this case (2.4.8) becomes

$$f_X(x) = \begin{cases} \frac{x^{n/2-1}e^{-x/2}}{\Gamma(n/2)2^{n/2}}, & \text{if } \ x > 0 \\ 0, & \text{otherwise.} \end{cases} \tag{2.4.9}$$

A random variable X having pdf given by (2.4.9) is said to have a **chi-square distribution with n degree of freedom**. The chi-square distribution has many important applications in statistical inference.

Definition 2.4.26. The **beta function** is defined as follows:

$$\beta(a, b) = \int_0^1 x^{a-1}(1-x)^{b-1}dx, \quad a, b > 0.$$

The following relationship between the beta function and the Gamma function is known.

$$\beta(a, b) = \frac{\Gamma(a)\Gamma(b)}{\Gamma(a+b)}. \tag{2.4.10}$$

Definition 2.4.27. A continuous random variable X is said to have a **beta distribution** with parameters $a > 0$ and $b > 0$, we write $X \sim B(a, b)$, if its pdf is of the form

$$f_X(x) = \begin{cases} \frac{x^{a-1}(1-x)^{b-1}}{\beta(a,b)}, & \text{if } \ 0 < x < 1 \\ 0, & \text{otherwise.} \end{cases} \tag{2.4.11}$$

Example 2.4.28. Let $X \sim B(a, b)$. Using the equality (2.4.10) and properties of the Gamma function, find $E[X]$ and $Var(X)$.

SOLUTION. Let n is a positive number. Then

$$\begin{aligned} E[X^n] &= \frac{1}{\beta(a,b)} \int_0^1 x^{n+a-1}(1-x)^{b-1}dx \\ &= \frac{\beta(n+a,b)}{\beta(a,b)} \\ &= \frac{\Gamma(n+a)\Gamma(a+b)}{\Gamma(a)\Gamma(n+a+b)}. \end{aligned} \tag{2.4.12}$$

Since $\Gamma(p) = (p-1)\Gamma(p-1)$, we have

$$E[X^n] = \frac{(n-1+a)\Gamma(n-1+a)\Gamma(a+b)}{\Gamma(a)(n-1+a+b)\Gamma(n-1+a+b)}. \tag{2.4.13}$$

Letting $n = 1$, we obtain

$$E[X] = \frac{a}{a+b}$$

Again applying the property of the Gamma function in the right hand side of (2.4.13), we have

$$E[X^n] = \frac{(n-1+a)(n-2+a)\Gamma(n-2+a)\Gamma(a+b)}{\Gamma(a)(n-1+a+b)(n-2+a+b)\Gamma(n-2+a+b)}.$$

Letting $n = 2$, we obtain

$$E[X^2] = \frac{a(a+1)}{(1+a+b)(a+b)}.$$

Thus

$$\begin{aligned} Var(X) &= \frac{a(a+1)}{(1+a+b)(a+b)} - \frac{a^2}{(a+b)^2} \\ &= \frac{ab}{(1+a+b)(a+b)^2}. \end{aligned}$$

Definition 2.4.29. A continuous random variable X is said to have a **Pareto distribution** with parameters $\alpha > 0$ and $\beta > 0$, we write $X \sim P(\alpha, \beta)$, if its pdf is of the form

$$f_X(x) = \begin{cases} \frac{\alpha\beta^\alpha}{x^{\alpha+1}}, & \text{if } x > \beta \\ 0, & \text{otherwise.} \end{cases} \tag{2.4.14}$$

Example 2.4.30. Let $X \sim P(\alpha, \beta)$. Find $E[X]$ and $Var(X)$.

SOLUTION. By the definition of the expectation of a continuous random variable, we have

$$\begin{aligned} E[X] &= \int_{-\infty}^{\infty} x f_X(x)dx = \int_{\beta}^{\infty} x \frac{\alpha\beta^\alpha}{x^{\alpha+1}} dx \\ &= \alpha\beta^\alpha \int_{\beta}^{\infty} \frac{1}{x^\alpha} dx = \alpha\beta^\alpha \left[-\frac{1}{(\alpha-1)x^{\alpha-1}} \right]_{\beta}^{\infty} \\ &= \frac{\alpha\beta}{\alpha-1} \end{aligned}$$

and

$$\begin{aligned} E[X^2] &= \int_{-\infty}^{\infty} x^2 f_X(x)dx = \int_{\beta}^{\infty} x^2 \frac{\alpha\beta^\alpha}{x^{\alpha+1}} dx \\ &= \alpha\beta^\alpha \int_{\beta}^{\infty} \frac{1}{x^{\alpha-1}} dx = \alpha\beta^\alpha \left[-\frac{1}{(\alpha-2)x^{\alpha-2}} \right]_{\beta}^{\infty} \\ &= \frac{\alpha\beta^2}{\alpha-2}. \end{aligned}$$

Therefore

$$\begin{aligned} Var(X) &= \frac{\alpha\beta^2}{\alpha-2} - \left(\frac{\alpha\beta}{\alpha-1}\right)^2 \\ &= \frac{\alpha\beta^2}{(\alpha-2)(\alpha-1)^2}. \end{aligned}$$

Definition 2.4.31. A continuous random variable X is said to have a **Weibull distribution** with parameters $\alpha > 0$ and $\beta > 0$, we write $X \sim W(\alpha, \beta)$, if its pdf is of the form

$$f_X(x) = \begin{cases} \alpha\beta x^{\beta-1} \exp\{-\alpha x^\beta\}, & \text{if } x > 0 \\ 0, & \text{otherwise.} \end{cases} \tag{2.4.15}$$

It is known that if $X \sim W(\alpha, \beta)$, then

$$E[X] = \alpha^{-1/\beta}\Gamma\left(\frac{1}{\beta}+1\right),$$

$$Var(X) = \alpha^{-2/\beta}\left\{\Gamma\left(\frac{2}{\beta}+1\right) - \left[\Gamma\left(\frac{1}{\beta}+1\right)\right]^2\right\}.$$

It is easy to check that if $X \sim W(\alpha, 1)$, then $X \sim E(\alpha)$.

Example 2.4.32. Find the failure rate $h_X(x)$ for $X \sim W(\alpha, \beta)$.

SOLUTION. Since the pdf $f_X(x)$ of X is given by

$$f_X(x) = \alpha\beta x^{\beta-1}\exp\{-\alpha x^\beta\}, \quad \text{for } x > 0,$$

the cumulative distribution function $F_X(x)$ is given by

$$\begin{aligned} F_X(x) = P(0 < X \le x) &= \int_0^x f_X(x)dx \\ &= \int_0^x \alpha\beta x^{\beta-1}\exp\{-\alpha x^\beta\}dx \\ &= \left[\exp\{-\alpha x^\beta\}\right]_0^x \\ &= 1 - \exp\{-\alpha x^\beta\}. \end{aligned}$$

Thus the survival function is

$$\begin{aligned} S_X(x) &= P(X > x) = 1 - F_X(x) \\ &= \exp\{-\alpha x^\beta\} \end{aligned}$$

and hence the failure rate $h_X(x)$ is given by

$$\begin{aligned} h_X(x) &= \frac{f_X(x)}{S_X(x)} \\ &= \frac{\alpha\beta x^{\beta-1}\exp\{-\alpha x^\beta\}}{\exp\{-\alpha x^\beta\}} \\ &= \alpha\beta x^{\beta-1}. \end{aligned}$$

Example 2.4.33. A company has 10 electric generators. These fail independently and each has common failure rate $h(t) = 10^{-4}t$ per day.

Find the expectation of the number of survivors among 10 electric generators, after 1,000 days.

SOLUTION. Let T be the lifetime of a single electric generator. Then survival function $S_T(t)$ is given by

$$\begin{aligned} S_T(t) &= 1 - F_T(t) = 1 - P(T \le t) \\ &= 1 - \int_0^t f_T(s)ds \\ &= 1 - \int_0^t h(s)S_T(s)ds. \end{aligned}$$

From this we have the following ordinary differential equation.

$$\begin{aligned} S_T'(t) &= -h(t)S_T(t) \\ S_T(0) &= 1. \end{aligned}$$

To solve this differential equation, rewrite as follows.

$$dS_T(t) = -h(t)S_T(t)dt$$

or

$$\frac{1}{S_T(t)}dS_T(t) = -h(t)dt.$$

Taking the definite integral from 0 to t in both side, we get

$$\ln S_T(t) - \ln S_T(0) = -\int_0^t h(s)ds.$$

Since $S_T(0) = 1$ and $\ln 1 = 0$, we have

$$S_T(t) = \exp\left\{-\int_0^t h(s)ds\right\}$$

Since $h(t) = 10^{-4}t$, we have

$$S_T(t) = \exp\left\{-\int_0^t 10^{-4}sds\right\} = \exp\left\{-10^{-4}\frac{t^2}{2}\right\}.$$

Thus each electric generator has a probability

$$\begin{aligned} S_T(1000) &= \exp\left\{-10^{-4}\frac{1000^2}{2}\right\} \\ &= \exp\left\{-10^{-4}\frac{10^6}{2}\right\} = e^{-50} \end{aligned}$$

of surviving the 1000 days. Then the expectation of the number of survivors among 10 electric generators is $10e^{-50}$.

Example 2.4.34. An electric device has a failure rate $h(t) = 2 \times 10^{-6}t$ per hour. Determine the probability that it will still be operational after 1,000 hours of service.

SOLUTION. From Example 2.4.32, we see that the failure rate $h(t) = 2 \times 10^{-6}t = 10^{-6} \times 2t^{2-1}$ is the failure rate of $T \sim W(10^{-6},\ 2)$. Thus, by Example 2.4.32, the survival function is given by

$$S_T(t) = \exp\{10^{-6}t^2\}$$

and hence

$$S_T(1000) = \exp\{10^{-6}1000^2\} = e^1 = 0.368.$$

CHAPTER 3

Joint and Conditional Distributions

3.1. Joint distributions

In many random experiments some observations involve the joint behavior of two or more random variables. For example, if a student's name is selected from an urn and s is an outcome, a pair $(X(s), Y(s))$ may be of interest, where $X(s)$ is the age of student s and $Y(s)$ is the height of student s in centimeters. Let X and Y be two random variable on the sample space of some experiment. Then the pair (X, Y) that assigns a point $(X(s), Y(s)) \in \mathbf{R}^2$ to each point $s \in S$ is called a two-dimensional **random vector**. Similarly, n-dimensional random vector can be defined. In Chapter 2, we saw that the probabilities of events involving a single random variable X can be expressed in terms of the distribution function $F_X(x)$. We also saw that these probabilities can be expressed as the sum over the pmf if the random variable is discrete, or the integral of the pdf if the random variable is continuous. We now extend these results to the two-dimensional random vector.

Definition 3.1.1. The function $F_{X,Y}(x, y)$, defined by

$$F_{X,Y}(x, y) = P(X \leq x, Y \leq y), \quad (x, y) \in \mathbf{R}^2,$$

is called the **joint distribution function** of the random vector (X, Y), or of X and Y.

Remark 3.1.2. By the same discussion in Section 2.1, it is easily shown that

(1) $0 \le F_{X,Y}(x, y) \le 1$.

(2) $\lim_{x,y\to\infty} F_{X,Y}(x, y) = 1$.

(3) $\lim_{x\to-\infty} F_{X,Y}(x, y) = 0 \quad \lim_{y\to-\infty} F_{X,Y}(x, y) = 0$.

(4) $F_{X,Y}(x, y)$ is a nondecreasing function of x and y.

(5) $F_{X,Y}(x, y)$ is right continuous.

(6) $\lim_{x\to\infty} F_{X,Y}(x, y) = F_Y(y), \quad \lim_{y\to\infty} F_{X,Y}(x, y) = F_X(x)$.

In (6), $F_X(x)$ and $F_Y(y)$ are called the **marginal distribution functions** of X and Y, respectively.

Let X and Y be two discrete random variables. Then the range $\mathcal{R}(X, Y)$ of the random vector (X, Y) consists of at most a countably infinite number of values (x_i, y_j), $i, j = 1, 2, \cdots$. A function $p_{X,Y}$ defined by

$$p_{X,Y}(x_i, y_j) = P(X = x_i, Y = y_j)$$

is called the **joint probability mass function** of X and Y. It is known that the joint pmf has the following two properties:

(a) $0 \le p_{X,Y}(x_i, y_j) \le 1$

(b) $\sum_{i=1}^{\infty} \sum_{j=1}^{\infty} p_{X,Y}(x_i, y_j) = 1$

and, conversely, any function on $\mathbf{R}^2$ satisfying these two properties is a joint pmf of some discrete random vector. It is easy to show that

$$F_{X,Y}(x, y) = \sum_{x_i \le x} \sum_{y_j \le y} p_{X,Y}(x_i, y_j).$$

Definition 3.1.3. (X, Y) is said to be a **continuous random vector** if there exists a nonnegative function $f_{X,Y}(x, y)$ such that

$$F_{X,Y}(x, y) = \int_{-\infty}^{x} \left(\int_{-\infty}^{y} f_{X,Y}(u, v)dv \right) du, \quad (x, y) \in \mathbf{R}^2,$$

Here $f_{X,Y}(x, y)$ is called the **joint probability density function** of X and Y.

Clearly the joint pdf $f_{X,Y}(x, y)$ of X and Y satisfies the following two properties:

(a) $f_{X,Y}(x, y) \geq 0$

(b) $\int_{-\infty}^{\infty} \int_{-\infty}^{\infty} f_{X,Y}(x, y)dxdy = 1$

and it is known that any function satisfying these two properties is the joint pdf of some continuous random vector. Consider a rectangular region $A = \{(x, y) \in \mathbf{R}^2 | \ a \leq x \leq b, c \leq y \leq d\}$. Then

$$\begin{aligned}
&P((X, Y) \in A) \\
&= P(a \leq X \leq b, c \leq Y \leq d) \\
&= F_{X,Y}(b, d) - F_{X,Y}(b, c) - F_{X,Y}(a, d) + F_{X,Y}(a, c) \\
&= \int_{-\infty}^{b} \int_{-\infty}^{d} f_{X,Y}(u, v)dvdu - \int_{-\infty}^{b} \int_{-\infty}^{c} f_{X,Y}(u, v)dvdu \\
&\qquad \int_{-\infty}^{a} \int_{-\infty}^{d} f_{X,Y}(u, v)dvdu + \int_{-\infty}^{a} \int_{-\infty}^{c} f_{X,Y}(u, v)dvdu \\
&= \int_{a}^{b} \int_{c}^{d} f_{X,Y}(u, v)dvdu
\end{aligned}$$

More general, if D is the **support** of $f_{X,Y}(x, y)$, that is, $f_{X,Y}(x, y) > 0$ on D and $= 0$ on D^c, and $A \in \mathbf{R}^2$ is a region in D, then

$$P((X, Y) \in A) = \int \int_{A} f_{X,Y}(x, y)dxdy, \tag{3.1.1}$$

which is a *double integral* over the region A. The double integral (3.1.1) is calculated by the following way; If the area A is given by

$$\begin{aligned} A &= \{(x,y) \in D|\ a \le x \le b, c(x) \le y \le d(x)\} \\ &= \{(x,y) \in D|\ a(y) \le x \le b(y), c \le y \le d\}, \end{aligned}$$

then (3.1.1) is same to the following *repeated integral.*

$$\begin{aligned} P((X,Y) \in A) &= \int_a^b \left(\int_{c(x)}^{d(x)} f_{X,Y}(x,y)dy \right) dx \\ &= \int_c^d \left(\int_{a(y)}^{b(y)} f_{X,Y}(x,y)dx \right) dy. \end{aligned}$$

Example 3.1.4. The joint pdf of two random variables X and Y is given by

$$f_{X,Y}(x,y) = \begin{cases} K(x+y), & 0 \le x, y \le 2 \\ 0, & \text{otherwise.} \end{cases}$$

(i) Find K. (ii) Evaluate $P(|X-Y| > 1)$.

SOLUTION. (i) By the definition of the joint pdf,

$$1 = \int_0^2 \int_0^2 K(x+y)dxdy = \int_0^2 K(2y+2)dy = 8K.$$

Hence $K = 1/8$.

(ii) Let D be the support of $f_{X,Y}(x,y)$ and

$$\begin{aligned} A &= \{(x,y) \in D|\ |x-y| > 1\} \\ &= \{(x,y) \in D|\ y > x+1\} \cup \{(x,y) \in D|\ y < x-1\} \\ &= \{(x,y) \in \mathbf{R}^2|\ 0 \le x \le 1, x+1 < y \le 2\} \\ &\qquad \cup \{(x,y) \in \mathbf{R}^2|\ 1 \le x \le 2, 0 \le y < x-1\} \\ &= A_1 \cup A_2. \end{aligned}$$

Therefore, from (3.1.1), we have

$$
\begin{aligned}
P(|X-Y|>1) &= P((X,Y)\in A) \\
&= \int\int_A f_{X,Y}(x,y)dxdy \\
&= \int\int_{A_1} f_{X,Y}(x,y)dxdy + \int\int_{A_2} f_{X,Y}(x,y)dxdy \\
&= 2\int\int_{A_1} f_{X,Y}(x,y)dxdy \\
&= 2\int_0^1 \left\{\int_{x+1}^2 \frac{1}{8}(x+y)dy\right\} dx \\
&= 2\int_0^1 \frac{1}{8}\left[xy+\frac{y^2}{2}\right]_{y=x+1}^{y=2} dx \\
&= 2\int_0^1 \left(-\frac{3x^2}{16}+\frac{3}{16}\right) dx \\
&= 2\left[-\frac{x^3}{16}+\frac{3x}{16}\right]_{x=0}^{x=1} \\
&= \frac{1}{4}.
\end{aligned}
$$

Remark 3.1.5. (1) In the continuous case, we have

$$\frac{\partial^2 F_{X,Y}(x,y)}{\partial x\partial y} = f_{X,Y}(x,y).$$

(2) If $h(x,y)$ is a two variable function, then

$$E[h(X,Y)] = \begin{cases} \sum_x \sum_y h(x,y)p_{X,Y}(x,y), \\ \int_{-\infty}^{\infty}\int_{-\infty}^{\infty} h(x,y)f_{X,Y}(x,y)dxdy \end{cases}$$

according as the random vector (X,Y) is discrete or continuous.
(3) If the support of $h(x,y)f_{X,Y}(x,y)$ is given by

$$
\begin{aligned}
D &= \left\{(x,y)\in \mathbf{R}^2 \mid \alpha\le x\le\beta, \gamma(x)\le y\le\delta(x)\right\} \\
&= \left\{(x,y)\in \mathbf{R}^2 \mid \alpha(y)\le x\le\beta(y), \gamma\le y\le\delta\right\},
\end{aligned}
$$

then

$$E[h(X,Y)] = \int_{\alpha}^{\beta}\left(\int_{\gamma(x)}^{\delta(x)} h(x,y)f_{X,Y}(x,y)dy\right)dx$$
$$= \int_{\gamma}^{\delta}\left(\int_{\alpha(y)}^{\beta(y)} h(x,y)f_{X,Y}(x,y)dx\right)dy.$$

Example 3.1.6. Let X and Y be two discrete random variables with the joint pmf $p_{X,Y}(x_i, y_j)$ which have the following values.

$y_j \setminus x_i$	-1	0	1
-1	1/18	3/18	3/18
0	1/18	0	3/18
1	4/18	1/18	2/18

Evaluate $E[X^2Y]$.

SOLUTION. By Remark 3.1.5, we have

$$\begin{aligned} E[X^2Y)] &= \sum_{x_i}\sum_{y_j} x_i^2 y_j p_{X,Y}(x_i,y_j) \\ &= (-1)^2 \times (-1) \times \frac{1}{18} + (-1)^2 \times 0 \times \frac{1}{18} + (-1)^2 \times 1 \times \frac{4}{18} \\ &\quad + 0 \times (-1) \times \frac{3}{18} + 0 \times 0 \times 0 + 0 \times 1 \times \frac{1}{18} \\ &\quad + 1^2 \times (-1) \times \frac{3}{18} + 1^2 \times 0 \times \frac{3}{18} + 1^2 \times 1 \times \frac{2}{18} \\ &= -\frac{1}{18} + \frac{4}{18} - \frac{3}{18} + \frac{2}{18} \\ &= \frac{1}{9}. \end{aligned}$$

Definition 3.1.7. In the case that X and Y are discrete,

$$p_X(x) = \sum_y p_{X,Y}(x,y) \quad \text{and} \quad p_Y(y) = \sum_x p_{X,Y}(x,y)$$

are called the **marginal pmf** of X and the marginal pmf of Y, respectively. Similarly, in the case that X and Y are continuous,

$$f_X(x) = \int_{-\infty}^{\infty} f_{X,Y}(x,y)dy \quad \text{and} \quad f_Y(y) = \int_{-\infty}^{\infty} f_{X,Y}(x,y)dx$$

are called the **marginal pdf** of X and the marginal pdf of Y, respectively.

Remark 3.1.8. Let $F_X(x)$ and $f_X(x)$ be the marginal distribution function and marginal pdf of X, respectively. Then we have

$$\begin{aligned} F_X(x) &= \lim_{y\to\infty} F_{X,Y}(x,y) \\ &= \lim_{y\to\infty} \int_{-\infty}^{x} \left(\int_{-\infty}^{y} f_{X,Y}(u,v)dv \right) du \\ &= \int_{-\infty}^{x} \left(\int_{-\infty}^{\infty} f_{X,Y}(u,v)dv \right) du \\ &= \int_{-\infty}^{x} f_X(u)du \end{aligned}$$

And so we obtain

$$\frac{\partial F_X(x)}{\partial x} = f_X(x).$$

Definition 3.1.9. We say that X and Y are **independent** if and only if

$$F_{X,Y}(x,y) = F_X(x)F_Y(y)$$

for all $(x,y) \in \mathbf{R}^2$.

Theorem 3.1.10. *A necessary and sufficient condition for continuous random variables X and Y to be independent is that*

$$f_{X,Y}(x,y) = f_X(x)f_Y(y)$$

PROOF. Let X and Y be independent. Then by Remark 3.1.8,

$$\begin{aligned} f_{X,Y}(x,y) &= \frac{\partial^2 F_{X,Y}(x,y)}{\partial x \partial y} \\ &= \frac{\partial^2 F_X(x)F_Y(y)}{\partial x \partial y} \\ &= \left(\frac{\partial F_X(x)}{\partial x}\right)\left(\frac{\partial F_Y(y)}{\partial y}\right) \\ &= f_X(x) f_Y(y). \end{aligned}$$

Conversely, if $f_{X,Y}(x,y) = f_X(x)f_Y(y)$, then by Remark 3.1.8 again, we see that

$$\begin{aligned} F_{X,Y}(x,y) &= \int_{-\infty}^{x}\left(\int_{-\infty}^{y} f_{X,Y}(u,v)dv\right)du \\ &= \int_{-\infty}^{x}\left(\int_{-\infty}^{y} f_X(u)f_Y(v)dv\right)du \\ &= \left(\int_{-\infty}^{x} f_X(u)du\right)\left(\int_{-\infty}^{y} f_Y(v)dv\right) \\ &= F_X(x)F_Y(y). \end{aligned}$$

Thus X and Y be independent. The proof is complete. □

Example 3.1.11. Are two random variables X and Y in Example 3.1.4 independent?

SOLUTION. To determine whether X and Y are independent, we first find their marginal pdf's and apply Theorem 3.1.10.

$$\begin{aligned} f_X(x) &= \int_0^2 \frac{1}{8}(x+y)dy \\ &= \frac{1}{8}\left[xy + \frac{y^2}{2}\right]_{y=0}^{y=2} \\ &= \begin{cases} \frac{1}{4}(x+1), & 0 \le x \le 2 \\ 0, & \text{otherwise} \end{cases} \end{aligned}$$

and since our joint pdf $f_{X,Y}(x,y)$ is symmetric as a function and on its support, we have

$$f_Y(y) = \begin{cases} \frac{1}{4}(y+1), & 0 \le y \le 2 \\ 0, & \text{otherwise.} \end{cases}$$

We can see $\frac{1}{8}(x+y) = f_{X,Y}(x,y) \neq f_X(x)f_Y(y) = \frac{1}{16}(x+1)(y+1)$ and so they are not independent random variables.

Remark 3.1.12. Let X and Y be independent. Then, by Remark 3.1.5(2) and Theorem 3.1.10, we see that for any functions g and h,

$$\begin{aligned} E[g(X)h(Y)] &= \int_{-\infty}^{\infty}\int_{-\infty}^{\infty} g(x)h(y)f_{X,Y}(x,y)dxdy \\ &= \int_{-\infty}^{\infty}\int_{-\infty}^{\infty} g(x)h(y)f_X(x)f_Y(y)dxdy \\ &= \left(\int_{-\infty}^{\infty} g(x)f_X(x)dx\right)\left(\int_{-\infty}^{\infty} h(y)f_Y(y)dy\right) \\ &= E[g(X)]E[h(Y)] \end{aligned}$$

In particular, if we take $g(z) = h(z) = z$, then

$$E[XY] = E[X]E[Y].$$

Definition 3.1.13. Let X and Y be two random variables.

$$Cov(X,Y) = E[(X - E[X])(Y - E[Y])] = E[XY] - E[X]E[Y]$$

is called the **covariance** between X and Y, and

$$\rho_{X,Y} = \frac{Cov(X,Y)}{\sqrt{Var(X)}\sqrt{Var(Y)}}$$

is called the **correlation coefficient** between X and Y.

Note that $Cor(X,X) = Var(X)$.

Remark 3.1.14. Let X, Y and Z be three random variables. Then for any real numbers a, b, c, we have

$$Cov(aX + bY, cZ) = abCov(X, Z) + acCov(X, Z).$$

In fact, by the definition of the covariance,

$$\begin{aligned}
&Cov(aX + bY, cZ)\\
&= E[(aX + bY)(cZ)] - E[aX + bY]E[cZ]\\
&= E[acXZ] + E[bcYZ] - E[aX]E[cZ] - E[bY]E[cZ]\\
&= ac\{E[XZ] - E[X]E[Z]\} + ac\{E[YZ] - E[Y]E[Z]\}\\
&= abCov(X, Z) + acCov(X, Z).
\end{aligned}$$

Example 3.1.15. Let X and Y be two discrete random variables with the joint pmf in Example 3.1.6 above. Find $\rho_{X,Y}$.

SOLUTION. Let us first find the marginal pmf $p_X(x_i)$ of X.

$$\begin{aligned}
p_X(-1) &= P(X = -1)\\
&= P(\{X = -1\} \cap S)\\
&= P(\{X = -1\} \cap [\{Y = -1\} \cup \{Y = 0\} \cup \{Y = 1\}])\\
&= P(X = -1, Y = -1) + P(X = -1, Y = 0) + P(X = -1, Y = 1)\\
&= \frac{1}{18} + \frac{1}{18} + \frac{4}{18} = \frac{3}{9}.
\end{aligned}$$

Similarly we have

$$p_X(0) = P(X = 0) = \frac{3}{18} + 0 + \frac{1}{18} = \frac{2}{9}$$

and

$$p_X(1) = P(X = 1) = \frac{3}{18} + \frac{3}{18} + \frac{2}{18} = \frac{4}{9},$$

so that

$$E[X] = \sum_{x_i} x_i p_X(x_i) = -1 \cdot \frac{3}{9} + 0 \cdot \frac{2}{9} + 1 \cdot \frac{4}{9} = \frac{1}{9}$$

and

$$E[X^2] = \sum_{x_i} x_i^2 p_X(x_i) = (-1)^2 \cdot \frac{3}{9} + 0 \cdot \frac{2}{9} + 1^2 \cdot \frac{4}{9} = \frac{7}{9}.$$

Thus

$$Var(X) = E[X^2] - E[X]^2 = \frac{7}{9} - \left(\frac{1}{9}\right)^2 = \frac{48}{81}.$$

By the similar method, the marginal pmf $p_Y(y_j)$ of Y is found to be

$$p_Y(-1) = P(Y = -1) = \frac{1}{18} + \frac{3}{18} + \frac{3}{18} = \frac{7}{18},$$

$$p_Y(0) = P(Y = 0) = \frac{1}{18} + 0 + \frac{3}{18} = \frac{4}{18}$$

and

$$p_Y(1) = P(Y = 1) = \frac{4}{18} + \frac{1}{18} + \frac{2}{18} = \frac{7}{18},$$

so that

$$E[Y] = \sum_{y_j} y_j p_Y(y_j) = -1 \cdot \frac{7}{18} + 0 \cdot \frac{4}{18} + 1 \cdot \frac{7}{18} = 0$$

and

$$E[Y^2] = \sum_{y_j} y_j^2 p_Y(y_j) = (-1)^2 \cdot \frac{7}{18} + 0 \cdot \frac{4}{18} + 1^2 \cdot \frac{7}{18} = \frac{7}{9}.$$

Thus

$$Var(Y) = E[Y^2] - E[Y]^2 = \frac{7}{9} - 0^2 = \frac{49}{81}.$$

Now we find $E[XY]$ by the same way to Example 3.1.6.

$$\begin{aligned}
E[XY)] &- \sum_{x_i} \sum_{y_j} x_i y_j p_{X,Y}(x_i, y_j) \\
&= (-1) \times (-1) \times \frac{1}{18} + (-1) \times 0 \times \frac{1}{18} + (-1) \times 1 \times \frac{4}{18} \\
&\quad + 0 \times (-1) \times \frac{3}{18} + 0 \times 0 \times 0 + 0 \times 1 \times \frac{1}{18} \\
&\quad + 1 \times (-1) \times \frac{3}{18} + 1 \times 0 \times \frac{3}{18} + 1 \times 1 \times \frac{2}{18} \\
&= -\frac{1}{18} - \frac{4}{18} - \frac{3}{18} + \frac{2}{18} \\
&= -\frac{2}{9}.
\end{aligned}$$

Therefore we have

$$
\begin{aligned}
\rho_{X,Y} &= \frac{E[XY]-E[X]E[Y]}{\sqrt{Var(X)Var(Y)}} \\
&= \frac{-\frac{2}{9}-\frac{1}{9}\cdot 0}{\sqrt{\frac{48}{81}\cdot\frac{49}{81}}} \\
&= -\frac{3\sqrt{3}}{14}.
\end{aligned}
$$

Remark 3.1.16. (1) From Remark 3.1.12 and Definition 3.1.13, we see that if X and Y are independent, then $Cov(X,Y)=0$ and so $\rho_{X,Y}=0$.
(2) For any X and Y, $|\rho_{X,Y}|\le 1$ as we see in Example 3.1.15. In fact, from Theorem 2.2.14, it holds

$$
\begin{aligned}
|Cov(X,Y)| &= |E[(X-E[X])(Y-E[Y])]| \\
&\le \sqrt{E[(X-E[X])^2}\sqrt{E[(Y-E[Y])^2} \\
&= \sqrt{Var(X)}\sqrt{Var(Y)}
\end{aligned}
$$

and this shows that

$$
|\rho_{X,Y}| = \frac{|Cov(X,Y)|}{\sqrt{Var(X)}\sqrt{Var(Y)}} \le 1.
$$

(3) It holds

$$
\begin{aligned}
Var(X+Y) &= E[(X+Y)^2]-E[X+Y]^2 \\
&= E[X^2+2XY+Y^2]-(E[X]+E[Y])^2 \\
&= E[X^2]-E[X]^2+E[Y^2]-E[Y]^2+2E[XY]-2E[X]E[Y] \\
&= Var(X)+Var(Y)+2Cov(X,Y).
\end{aligned}
$$

From this we see that $Var(X+Y)=Var(X)+Var(Y)$ if X and Y are independent, and that $Var(2X)=4Var(X)$ if we take $X=Y$. In more general,

$$
\begin{aligned}
Var(aX+bY+c) &= Var(aX+bY) \\
&= Var(aX)+Var(bY)+2Cov(aX,bY) \\
&= a^2Var(X)+b^2Var(Y)+2abCov(X,Y).
\end{aligned}
$$

Definition 3.1.17. A continuous random vector (X, Y) is said to have a uniform distribution on a region $D \subset \mathbf{R}^2$ and write $(X, Y) \sim U(D)$ if the joint probability density function of X and Y is of the form

$$f_{X,Y}(x, y) = \begin{cases} \frac{1}{\text{area of } D}, & \text{for} \quad (x, y) \in D \\ 0, & \text{otherwise.} \end{cases}$$

Example 3.1.18. Let $(X, Y) \sim U(D)$, where $D = \{(x, y) |\ 0 \leq y \leq x \leq a\}$. Find $P(Y \geq X/2)$.

SOLUTION. Since $(X, Y) \sim U(D)$, X and Y have the joint pdf of the following form

$$\begin{aligned} f_{X,Y}(x, y) &= \begin{cases} \frac{1}{\text{area of } D}, & \text{for} \quad (x, y) \in D \\ 0, & \text{otherwise} \end{cases} \\ &= \begin{cases} \frac{2}{a^2}, & \text{for} \quad (x, y) \in D \\ 0, & \text{otherwise.} \end{cases} \end{aligned}$$

Since

$$P(Y \geq X/2) = P((X, Y) \in A),$$

where

$$\begin{aligned} A &= \{(x, y) \in D |\ y \geq \frac{x}{2}\} \\ &= \{(x, y) |\ 0 \leq x \leq a, \frac{1}{2}x < y \leq x\}, \end{aligned}$$

it follows that

$$\begin{aligned} P(Y \geq X/2) &= \int\int_A f_{X,Y}(x, y) dx dy = \int_0^a \left[\int_{x/2}^x \frac{2}{a^2} dy \right] dx \\ &= \int_0^a \left[\frac{2}{a^2} y \right]_{x/2}^x dx = \int_0^a \frac{2}{a^2} \left(x - \frac{x}{2} \right) dx \\ &= \int_0^a \frac{x}{a^2} dx = \left[\frac{x^2}{2a^2} \right]_0^a = \frac{1}{2}. \end{aligned}$$

Definition 3.1.19. A continuous random variables $X_1, X_2, \cdots, X_n$ is said to have a **multivariate normal distribution** with mean vector $\vec{\mu}$ and covariance matrix Σ and write $X \sim LN(\vec{\mu}, \Sigma)$ if their joint pdf is of the form

$$f(\mathbf{x}) = \frac{1}{(2\pi)^{n/2}|\Sigma|^{1/2}} \exp\left\{-\frac{1}{2}(\mathbf{x}-\vec{\mu})'\Sigma^{-1}(\mathbf{x}-\vec{\mu})\right\}, \tag{3.1.2}$$

where

$$\mathbf{x} = (x_1, x_2, \cdots, x_n)',$$

$$\vec{\mu} = (\mu_1, \mu_2, \cdots, \mu_n)'$$

and

$$\begin{aligned}\Sigma &= E[(\mathbf{x}-\vec{\mu})(\mathbf{x}-\vec{\mu})'] \\ &= \begin{pmatrix} \sigma_1^2 & \sigma_{12} & \cdots & \sigma_{1n} \\ \sigma_{21} & \sigma_2^2 & \cdots & \sigma_{2n} \\ \cdots & \cdots & \cdots & \cdots \\ \sigma_{n1} & \sigma_{n2} & \cdots & \sigma_n^2 \end{pmatrix}\end{aligned}$$

Example 3.1.20. Derive the bivariate normal density function of two random variables X_1 and X_2 from the equality (3.1.2).

SOLUTION. Let the correlation coefficient of X_1 and X_2 be ρ. Then, the covariance matrix becomes

$$\Sigma = \begin{pmatrix} \sigma_1^2 & \rho\sigma_1\sigma_2 \\ \rho\sigma_1\sigma_2 & \sigma_2^2 \end{pmatrix}$$

and hence, we have

$$|\Sigma| = (1-\rho^2)\sigma_1^2\sigma_2^2.$$

The inverse of the covariance matrix can be obtained as

$$\Sigma^{-1} = \frac{1}{1-\rho^2}\begin{pmatrix} 1/\sigma_1^2 & -\rho/(\sigma_1\sigma_2) \\ -\rho/(\sigma_1\sigma_2) & 1/\sigma_2^2 \end{pmatrix}$$

By inserting $|\Sigma|$ and Σ^{-1} into the equality (3.1.2), we can drive the joint pdf of X_1 and X_2 given by, for $-\infty < x_1, x_2 < \infty$,

$$f_{X_1,X_2}(x_1,x_2) = \frac{1}{2\pi\sigma_1\sigma_2\sqrt{1-\rho^2}} \exp\left\{-\frac{1}{2(1-\rho^2)}\left[\left(\frac{x_1-\mu_1}{\sigma_1}\right)^2 \right.\right.$$
$$\left.\left. -2\rho\left(\frac{x_1-\mu_1}{\sigma_1}\right)\left(\frac{x_2-\mu_2}{\sigma_2}\right)+\left(\frac{x_2-\mu_2}{\sigma_2}\right)^2\right]\right\}$$

where μ_1 and μ_2 are the means, σ_1^2 and σ_2^2 are the variances of X_1 and X_2, respectively, while ρ is the correlation coefficient of X_1 and X_2.

3.2. Conditional distributions

We see that many random variables of practical interest are not independent. For example, the output Y of a communication channel must be dependent on the input X in order to convey information. In this section we are interested in computing the probability of events concerning the random variable Y given that we know that $X = x$. We are also interested in the expected value of Y given that $X = x$.

Definition 3.2.1. The function $F_{X|Y}(x|Y = y)$, defined by

$$F_{X|Y}(x|Y = y) = \lim_{h\to 0} P(X \le x | y \le Y \le y+h) \tag{3.2.1}$$

for $-\infty < x < \infty$, is called the **conditional distribution function** of the random variable X given that $Y = y$. If (X, Y) is a discrete random vector, then the right hand side of (3.2.1) equals $P(X \le x|Y = y)$. In the case that (X, Y) is a continuous random vector, we will write the right hand side of (3.2.1) by $P(X \le x|Y = y)$ also.

Let X and Y are discrete random variables and be given an event $\{Y = y\}$. A function $p_{X|Y}(\cdot|Y = y_j)$ defined by

$$p_{X|Y}(x_i|Y = y_j) = P(X = x_i|Y = y_j) = \frac{p_{X,Y}(x_i, y_j)}{p_Y(y_j)}$$

is called the **conditional probability mass function** of X given that $Y = y_j$. In the discrete case, we can see that

(a) $0 \leq p_{X|Y}(x_i|Y = y_j) \leq 1$

(b) $\sum_{i=1}^{\infty} p_{X|Y}(x_i|Y = y_j) = 1$

and

(c)$F_{X|Y}(x|Y = y_j) = \sum_{x_i \leq x} p_{X|Y}(x_i|Y = y_j)$.

In fact, the property (b) is proved from the definition of the conditional pmf and the property $\sum_{i=1}^{\infty} p_{X,Y}(x_i, y_j) = p_Y(y_j)$.

Definition 3.2.2. Let (X, Y) be a continuous random vector. A non-negative function $f_{X|Y}(\cdot|Y = y)$ satisfying

$$F_{X|Y}(x|Y = y) = \int_{-\infty}^{x} f_{X|Y}(u|Y = y)du, \quad -\infty < x < \infty$$

is called the **conditional probability density function** of the random variable X given that $Y = y$.

Theorem 3.2.3. *Let X and Y denote two continuous random variables. Then the joint pdf $f_{X,Y}(x, y)$ and the conditional pdf $f_{X|Y}(x|Y = y)$ of them have the following relation.*

$$f_{X|Y}(x|Y = y) = \frac{f_{X,Y}(x, y)}{f_Y(y)}.$$

PROOF. By Definition 3.2.1 and the mean value theorem, we have

$$
\begin{aligned}
F_{X|Y}(x|Y=y) &= \lim_{h\to 0} P(X \le x | y \le Y \le y+h) \\
&= \lim_{h\to 0} \frac{P(X \le x, y \le Y \le y+h)}{P(y \le Y \le y+h)} \\
&= \lim_{h\to 0} \frac{\int_{-\infty}^{x}\int_{y}^{y+h} f_{X,Y}(u,v)\,dv\,du}{\int_{y}^{y+h} f_Y(v)\,dv} \\
&= \lim_{h\to 0} \frac{\int_{-\infty}^{x} h f_{X,Y}(u,y_1)\,du}{h f_Y(y_2)} \qquad \text{for some} \quad y_1, y_2 \in (y, y+h) \\
&= \int_{-\infty}^{x} \frac{f_{X,Y}(u,y)}{f_Y(y)}\,du
\end{aligned}
$$

Combining this with Definition 3.2.2, the proof is completed. □

In the continuous case, we can see easily that

(a) $0 \le f_{X|Y}(u|Y=y)$

(b) $\int_{-\infty}^{\infty} f_{X|Y}(x|Y=y)\,dx = 1.$

In fact, the property (b) is proved by Theorem 3.2.3 and the definition of the marginal pdf.

Example 3.2.4. The joint pdf of two random variables X and Y is given by

$$
f_{X,Y}(x,y) = \begin{cases} \frac{4(x+y)}{5x^3}, & 0 \le y \le 1, 1 \le x < \infty \\ 0, & \text{otherwise.} \end{cases}
$$

(i) Find $P(0 < Y < \frac{1}{2}|X=2)$. (ii) For what value of K is it true that $P(0 < Y < \frac{1}{2}|X > K) = 5/16$?

SOLUTION. (i) To applying Definition 3.2.1 and Definition 3.2.2, we have to find $f_{Y|X}(y|X=x)$. To do this, we first find $f_X(x)$.

$$\begin{aligned} f_X(x) &= \int_0^1 \frac{4(x+y)}{5x^3}dy \\ &= \frac{4}{5}\left[\frac{y}{x^2}+\frac{y^2}{2x^3}\right]_{y=0}^{y=1} \\ &= \frac{4}{5}\left(\frac{1}{x^2}+\frac{1}{2x^3}\right), \quad 1 \le x < \infty. \end{aligned}$$

By Theorem 3.2.3,

$$f_{Y|X}(y|X=x) = \frac{4(x+y)/(5x^3)}{\frac{4}{5}(1/x^2+1/(2x^3))} = \frac{2(x+y)}{2x+1}, \quad 0 \le y \le 1.$$

Thus, by Definition 3.2.1 and Definition 3.2.2,

$$\begin{aligned} P\left(0<Y<\frac{1}{2}|X=2\right) &= \int_0^{1/2} f_{Y|X}(y|X=2)dy \\ &= \int_0^{1/2} \frac{2(2+y)}{5}dy \\ &= \frac{2}{5}\left[2y+\frac{y^2}{2}\right]_0^{1/2} \\ &= \frac{9}{20}. \end{aligned}$$

(ii) By the definition of the conditional probability,

$$P\left(0<Y<\frac{1}{2}|X>K\right) = \frac{P(0<Y<\frac{1}{2}, X>K)}{P(X>K)}.$$

Since

$$\begin{aligned} P(X>K) &= \int_K^\infty \frac{4}{5}\left(\frac{1}{x^2}+\frac{1}{2x^3}\right)dx \\ &= \frac{4}{5}\left(\frac{1}{K}+\frac{1}{4K^2}\right) \end{aligned}$$

and

$$\begin{aligned}
P(0 < Y < \frac{1}{2}, X > K) &= \int_K^\infty \int_0^{1/2} \frac{4(x+y)}{5x^3} dy dx \\
&= \frac{4}{5} \int_K^\infty \left[\frac{y}{x^2} + \frac{y^2}{2x^3} \right]_{y=0}^{y=1/2} dx \\
&= \frac{4}{5} \int_K^\infty \left(\frac{1}{2x^2} + \frac{1}{8x^3} \right) dx \\
&= \frac{4}{5} \left(\frac{1}{2K} + \frac{1}{16K^2} \right),
\end{aligned}$$

it holds

$$\begin{aligned}
\frac{5}{16} &= P(0 < Y < \frac{1}{2} | X > K) \\
&= \frac{\frac{4}{5}[1/(2K) + 1/(16K^2)]}{\frac{4}{5}[1/K + 1/(4K^2)]} \\
&= \frac{8K+1}{16K+4}.
\end{aligned}$$

This implies that $K = 1/12$.

Example 3.2.5. Let X be a continuous random variable with pdf

$$f_X(x) = \begin{cases} 2x, & 0 < x < 1 \\ 0, & \text{otherwise.} \end{cases}$$

and Y be a continuous random variable such that the conditional distribution function $F_{Y|X}(y|X = x)$ of Y given that $X = x$ is uniform on the interval $(0, x)$. (i) Find $Var(Y)$. (ii) Find $\rho_{X,Y}$.

SOLUTION. (i) Since $f_X(x) = 2x$ for $0 < x < 1$ and $f_{Y|X}(y|X = x) = 1/x$ for $0 < y < x$, we have

$$f_{X,Y}(x, y) = f_X(x) f_{Y|X}(y|X = x) = 2x\frac{1}{x} = 2$$

for $(x, y) \in D = \{(x, y)|\ 0 < x < 1, 0 < y < x\}$. Note that $D = \{(x, y)|\ 0 < y < 1, y < x < 1\}$. By the definition of the marginal pdf,

$$\begin{aligned} f_Y(y) &= \int_{-\infty}^{\infty} f_{X,Y}(x, y)dx \\ &= \int_D 2dx = \int_y^1 2dx \\ &= 2(1-y) \end{aligned}$$

for $0 < y < 1$ and $f_Y(y) = 0$, otherwise. This gives

$$\begin{aligned} E[Y] &= \int_0^1 y \cdot 2(1-y)dy \\ &= 2\int_0^1 (y - y^2)dy \\ &= 2\left[\frac{y^2}{2} - \frac{y^3}{3}\right]_0^1 \\ &= \frac{1}{3} \end{aligned}$$

and

$$\begin{aligned} E[Y^2] &= \int_0^1 y^2 \cdot 2(1-y)dy \\ &= 2\int_0^1 (y^2 - y^3)dy \\ &= 2\left[\frac{y^3}{3} - \frac{y^4}{4}\right]_0^1 \\ &= \frac{1}{6}. \end{aligned}$$

Thus we have

$$Var(Y) = E[Y^2] - E[Y]^2 = \frac{1}{6} - \left(\frac{1}{3}\right)^2 = \frac{1}{18}.$$

(ii) Since $f_X(x) = 2x$ for $0 < x < 1$, we have

$$E[X] = \int_0^1 x \cdot 2xdx = \frac{2}{3}$$

and

$$E[X^2] = \int_0^1 x^2 \cdot 2xdx = \frac{1}{2}.$$

Thus

$$Var(X) = E[X^2] - E[X]^2 = \frac{1}{2} - \left(\frac{2}{3}\right)^2 = \frac{1}{18}.$$

Now we get

$$\begin{aligned} E[XY] &= \int\int_D xy f_{X,Y}(x,y)dxdy \\ &= \int_0^1 \left(\int_0^x 2xydy\right) dx \\ &= \int_0^1 x^3 dx \\ &= \frac{1}{4}. \end{aligned}$$

By the definition of $\rho_{X,Y}$, we have

$$\begin{aligned} \rho_{X,Y} &= \frac{E[XY] - E[X]E[Y]}{\sqrt{Var(X)Var(Y)}} \\ &= \frac{\frac{1}{4} - \frac{2}{3}\cdot\frac{1}{3}}{\sqrt{\frac{1}{18}\cdot\frac{1}{18}}} = \frac{\frac{1}{4} - \frac{2}{9}}{\frac{1}{18}} \\ &= \frac{1}{2}. \end{aligned}$$

Definition 3.2.6. The **conditional expectation** of X given that $Y = y$ is a real number defined by

$$E[X|Y = y] = \sum_{x_i} x_i p_{X|Y}(x_i|Y = y)$$

in the discrete case, and

$$E[X|Y = y] = \int_{-\infty}^{\infty} x dF_{X|Y}(x|Y = y) = \int_{-\infty}^{\infty} x f_{X|Y}(x|Y = y)dx$$

in the continuous case.

Definition 3.2.7. A random variable $E[X|Y]$ defined by

$$E[X|Y](s) = E[X|Y = y] \quad \text{if} \quad Y(s) = y$$

is called the **conditional expectation** of X given Y.

Theorem 3.2.8. *For any random vector (X, Y), it holds that*

$$E[E[X|Y]] = E[X].$$

PROOF. In the discrete case, since $E[X|Y] = E[X|Y = y_j]$ with probability $P(Y = y_j) = p_Y(y_j)$, we have

$$\begin{aligned}
E[E[X|Y]] &= \sum_{y_j} E[X|Y = y_j] p_Y(y_j) \\
&= \sum_{y_j} \sum_{x_i} x_i p_{X|Y}(x_i|Y = y_j) p_Y(y_j) \\
&= \sum_{y_j} \sum_{x_i} x_i \frac{p_{X,Y}(x_i, y_j)}{p_Y(y_j)} p_Y(y_j) \\
&= \sum_{x_i} x_i \sum_{y_j} p_{X,Y}(x_i, y_j) \\
&= \sum_{x_i} x_i p_X(x_i) \\
&= E[X].
\end{aligned}$$

Now we prove the theorem in the continuous case. Since

$$E[X|Y](s) = E[X|Y = y] \quad \text{if} \quad Y(s) = y,$$

from Theorem 3.2.3, we have

$$\begin{aligned}
E[E[X|Y]] &= \int_{-\infty}^{\infty} E[X|Y = y] f_Y(y) dy \\
&= \int_{-\infty}^{\infty} \left\{ \int_{-\infty}^{\infty} x f_{X|Y}(x|Y = y) dx \right\} f_Y(y) dy \\
&= \int_{-\infty}^{\infty} \left\{ \int_{-\infty}^{\infty} x \frac{f_{X,Y}(x, y)}{f_Y(y)} dx \right\} f_Y(y) dy \\
&= \int_{-\infty}^{\infty} x \left\{ \int_{-\infty}^{\infty} f_{X,Y}(x, y) dy \right\} dx \\
&= \int_{-\infty}^{\infty} x f_X(x) dx \\
&= E[X].
\end{aligned}$$

The proof is complete. □

Definition 3.2.9. (i) The **conditional variance** of X given that $Y = y$ is a real number defined by

$$Var(X|Y = y) = E[X^2|Y = y] - E[X|Y = y]^2.$$

(ii) A random variable $Var(X|Y)$ defined by

$$Var(X|Y) = E[X^2|Y] - E[X|Y]^2$$

is called the **conditional variance** of X given Y.

Theorem 3.2.10. *It holds*

$$Var(X) = Var(E[X|Y]) + E[Var(X|Y)]. \tag{3.2.2}$$

PROOF. The right hand side of the equality (3.2.2) is equal to

$$\begin{aligned}
&E[E[X|Y]^2] - E[E[X|Y]]^2 + E[E[X^2|Y] - E[X|Y]^2]\\
&= E[E[X^2|Y]] - E[E[X|Y]]^2\\
&= E[X^2] - E[X]^2\\
&= Var(X).
\end{aligned}$$

The proof is complete. □

Example 3.2.11. Let X and Y be two continuous random variables described in Example 3.2.5. By using Theorem 3.2.10, find $Var(Y)$.

SOLUTION. We are given $f_X(x) = 2x$ for $0 < x < 1$ and $f_{Y|X}(y|X = x) = 1/x$ for $0 < y < x$. Thus

$$\begin{aligned}
E[Y|X = x] &= \int_{\infty}^{\infty} y f_{Y|X}(y|X = x)dy\\
&= \int_0^x y \cdot \frac{1}{x} dy\\
&= \frac{x}{2},
\end{aligned}$$

so that $E[Y|X] = X/2$. Similarly,

$$\begin{aligned}E[Y^2|X=x] &= \int_{-\infty}^{\infty} y^2 f_{Y|X}(y|X=x)dy \\ &= \int_0^x y^2 \cdot \frac{1}{x}dy \\ &= \frac{x^2}{3},\end{aligned}$$

so that $E[Y^2|X] = X^2/3$. Thus, by Definition 3.2.9,

$$Var(X|Y) = E[X^2|Y] - E[X|Y]^2 = \frac{X^2}{3} - \left(\frac{X}{2}\right)^2 = \frac{X^2}{12}.$$

and hence

$$E[Var(X|Y)] = E\left[\frac{X^2}{12}\right] = \int_0^1 \frac{x^2}{12} \cdot 2xdx = \frac{1}{24}.$$

And it holds

$$\begin{aligned}Var(E[Y|X]) &= Var\left(\frac{X}{2}\right) \\ &= \frac{1}{4}Var(X) = \frac{1}{4}(E[X^2] - E[X]^2) \\ &= \frac{1}{4}\left\{\int_0^1 x^2 \cdot 2xdx - \left(\int_0^1 x \cdot 2xdx\right)^2\right\} \\ &= \frac{1}{4}\left\{\frac{1}{2} - \left(\frac{2}{3}\right)^2\right\} \\ &= \frac{1}{72}.\end{aligned}$$

By Theorem 3.2.10,

$$\begin{aligned}Var(X) &= Var(E[X|Y]) + E[Var(X|Y)] \\ &= \frac{1}{24} + \frac{1}{72} \\ &= \frac{1}{18}.\end{aligned}$$

Example 3.2.12. ([3]) Let X and Y denote the values of two stocks at the end of a five-year period. X is uniformly distributed on the interval

$(0, 12)$. Given $X = x$, Y is uniformly distributed on the interval $(0, x)$. Determine $Cov(X, Y)$ according to this model.

SOLUTION. By our assumption, we have $f_{Y|X}(y|X = x) = 1/x$ for $0 < y < x$ and $f_X(x) = 1/12$ for $0 < x < 12$. So we have

$$f_{X,Y}(x, y) = f_{Y|X}(y|X = x)f_X(x) = \frac{1}{12x}$$

for $0 < y < x < 12$. The marginal density of Y is

$$f_Y(y) = \int_{\infty}^{\infty} f_{X,Y}(x, y)dx = \int_y^{12} \frac{1}{12x}dx = \frac{1}{12}\ln\left(\frac{12}{y}\right)$$

for $0 < y < 12$. By the definition of the covariance,

$$Cov(X, Y) = E[XY] - E[X]E[Y].$$

Since X has a uniformly distributed on the interval $(0, 12)$, $E[X] = 12/2 = 6$. And we have

$$E[Y] = \int_0^{12} yf_Y(y)dy = \int_0^{12} y\frac{1}{12}\ln\left(\frac{12}{y}\right)dy.$$

Because it is not easy to calculate the definite integral in the right hand side, we use the "double expectation" approach, $E[Y] = E[E[Y|X]]$ and $E[XY] = E[E[XY|X]]$. Since the conditional random variable Y given $X = x$ is uniformly distributed on the interval $(0, x)$, we have $E[Y|X = x] = x/2$ and hence $E[Y|X] = X/2$. So

$$E[Y] = E[E[Y|X]] = E\left[\frac{X}{2}\right] = \frac{6}{2} = 3.$$

Also $E[XY] = E[E[XY|X]]$. But

$$E[XY|X] = XE[Y|X] = X \cdot \frac{X}{2} = \frac{X^2}{2}.$$

Then

$$E[XY] = E[E[XY|X]] = E\left[\frac{X^2}{2}\right] = \frac{1}{2}\int_0^{12} x^2\left(\frac{1}{12}\right)dx = 24.$$

Finally, $Cov(X, Y) = 24 - (6)(3) = 6.$

Alternatively, we can find $E[X]$, $E[Y]$ and $E[XY]$ by using only $f_{X,Y}(x,y)$ from Remark 3.1.5(2). To do this, let $D = \{(x,y)|0 < x < 12,\ 0 < y < x\}$. Then we have

$$E[X] = \int\int_D x f_{X,Y}(x,y)dxdy = \int_0^{12}\left[\int_0^x x\frac{1}{12x}dx\right]dy = 6,$$

$$E[Y] = \int\int_D y f_{X,Y}(x,y)dxdy = \int_0^{12}\left[\int_0^x y\frac{1}{12x}dx\right]dy = 3$$

and

$$E[XY] = \int\int_D xy f_{X,Y}(x,y)dxdy = \int_0^{12}\left[\int_0^x xy\frac{1}{12x}dx\right]dy = 24.$$

Therefore, it follows

$$Cov(X,Y) = E[XY] - E[X]E[Y] = 24 - (6)(3) = 6.$$

Remark 3.2.13. (1) If X and Y are two discrete random variables, then, by the theorem of total probability, we have

$$P(X \le x) = \sum_{k=1}^{\infty} P(X \le x|Y = k)p_Y(k).$$

(2) Let X and Y be two continuous random variables. From Theorem 3.2.3, we have

$$f_{X,Y}(x,y) = f_{X|Y}(x|Y = y)f_Y(y)$$

and so

$$\begin{aligned}\int_{-\infty}^{x} f_{X,Y}(u,y)du &= f_Y(y)\int_{-\infty}^{x} f_{X|Y}(u|Y = y)du \\ &= f_Y(y)F_{X|Y}(x|Y = y) \\ &= f_Y(y)P(X \le x|Y = y).\end{aligned}$$

Thus

$$\begin{aligned}P(X \le x) =& P(X \le x, -\infty < y < \infty) \\ =& \int_{-\infty}^{\infty}\left(\int_{-\infty}^{x} f_{X,Y}(u,y)du\right)dy \\ =& \int_{-\infty}^{\infty} P(X \le x|Y = y)f_Y(y)dy.\end{aligned}$$

Similarly, we have

$$P(X > x) = \int_{-\infty}^{\infty} P(X > x|Y = y) f_Y(y)dy.$$

Example 3.2.14. ([3]) An insurance policy is written to cover a loss X where X has density function

$$f_X(x) = \begin{cases} 3x^2/8, & 0 \le x \le 2 \\ 0, & \text{otherwise.} \end{cases}$$

The time (in hours) T to process a claim of size x, where $0 \le x \le 2$, is uniformly distributed on the interval from x to $2x$. Calculate the probability that a randomly chosen claim on this policy is processed in three hours or more.

SOLUTION. Since the distribution given claim amount $X = x$ is uniform on the interval $(x, 2x)$, the conditional pdf of T given that $X = x$ is $f_{T|X}(t|X = x) = 1/x$ for $x < t < 2x$. The pdf of X is $f_X(x) = 3x^2/8$ for $0 \le x \le 2$. And

$$\begin{aligned} P(T > 3|X = x) &= \int_3^{\infty} f_{T|X}(t|X = x)dt \\ &= \begin{cases} \int_3^{\infty} 0dt = 0, & \text{if } \ 2x \le 3, \quad \text{i.e.,} \quad x \le 1.5 \\ \int_3^{\infty} \frac{1}{x}dt = \frac{2x-3}{x} & \text{if } \ 3 < 2x \le 4 \quad \text{i.e.,} \quad 1.5 < x \le 2. \end{cases} \end{aligned}$$

Then, by Remark 3.2.13(2), we have

$$\begin{aligned} P(T > 3) &= \int_{1.5}^{2} P(T > 3|X = x) f_X(x)dx \\ &= \int_{1.5}^{2} \frac{2x-3}{x} \cdot \frac{3x^2}{8} dx = \frac{11}{64}. \end{aligned}$$

Definition 3.2.15. The **joint moment generating function** $M_{X,Y}(u, v)$ of X and Y is defined by

$$M_{X,Y}(u, v) = E[e^{uX+vY}].$$

The joint mgf $M_{X,Y}(u,v)$ has the following properties.

Theorem 3.2.16. *(i)* $M_{X,Y}(u,0) = M_X(u)$, $M_{X,Y}(0,v) = M_Y(v)$.
(ii) $\frac{\partial}{\partial u} M_{X,Y}(u,v)|_{u=v=0} = E[X]$ *and* $\frac{\partial}{\partial v} M_{X,Y}(u,v)|_{u=v=0} = E[Y]$.
(iii) $\frac{\partial^{m+n}}{\partial^m u \partial^n v} M_{X,Y}(u,v)|_{u=v=0} = E[X^m Y^n]$.
(iv) X and Y are independent if and only if

$$M_{X,Y}(u,v) = M_X(u)M_Y(v) \quad \textit{for all} \quad u,v \in \mathbf{R}.$$

PROOF. The property (i) is clear. (ii) is a special case of (iii). Let us prove (iii) in the continuous case. The proof in the discrete case is similar. By Remark 3.1.5(2),

$$\begin{aligned}\frac{\partial^{m+n}}{\partial^m u \partial^n v} M_{X,Y}(u,v) &= \frac{\partial^{m+n}}{\partial^m u \partial^n v} \int_{-\infty}^{\infty} \int_{-\infty}^{\infty} e^{ux+vy} f_{X,Y}(x,y)dxdy \\ &= \int_{-\infty}^{\infty} \int_{-\infty}^{\infty} x^m y^n e^{ux+vy} f_{X,Y}(x,y)dxdy.\end{aligned}$$

Thus, by Remark 3.1.5(2) again, we have

$$\begin{aligned}\frac{\partial^{m+n}}{\partial^m u \partial^n v} M_{X,Y}(u,v)|_{u=v=0} &= \int_{-\infty}^{\infty} \int_{-\infty}^{\infty} x^m y^n f_{X,Y}(x,y)dxdy. \\ &= E[X^m Y^n].\end{aligned}$$

Finally, we prove (iv) in the continuous case. Let X and Y be independent. Then

$$M_{X,Y}(u,v) = E[e^{uX+vY}] = E[e^{uX}]E[e^{vY}] = M_X(u)M_Y(v).$$

Conversely, if

$$M_{X,Y}(u,v) = M_X(u)M_Y(v),$$

then

$$\int_{-\infty}^{\infty} \int_{-\infty}^{\infty} e^{ux+vy} f_{X,Y}(x,y)dxdy = \int_{-\infty}^{\infty} e^{ux} f_X(x)dx \int_{-\infty}^{\infty} e^{vy} f_Y(y)dy,$$

that is,

$$\int_{-\infty}^{\infty} \int_{-\infty}^{\infty} e^{ux+vy} f_{X,Y}(x,y)dxdy = \int_{-\infty}^{\infty} \int_{-\infty}^{\infty} e^{ux+vy} f_X(x) f_Y(y)dxdy.$$

By the uniqueness of the joint mgf, we must have

$$f_{X,Y}(x,y) = f_X(x)f_Y(y) \quad \text{for all}(x,y) \in \mathbf{R}^2.$$

This implies that X and Y are independent. A similar proof is given in the discrete case. The proof is complete. □

Example 3.2.17. Let X and Y be two continuous random variables with joint pdf

$$f_{X,Y}(x,y) = \begin{cases} e^{-x-y}, & 0 < x < \infty, 0 < y < \infty \\ 0, & \text{otherwise.} \end{cases}$$

Find the joint mgf $M_{X,Y}(u,v)$ and the covariance $Cov(X,Y)$.

SOLUTION. By the definition of the joint mgf,

$$\begin{aligned} M_{X,Y}(u,v) &= E[e^{uX+vY}] \\ &= \int_0^\infty \int_0^\infty e^{ux+vy} e^{-x-y} dx dy. \\ &= \frac{1}{1-u} \cdot \frac{1}{1-v} \end{aligned}$$

for $u, v < 1$. Hence

$$E[X] = \frac{\partial M(0,0)}{\partial u} = \frac{1}{(1-u)^2} \cdot \frac{1}{1-v}|_{u=v=0} = 1,$$

$$E[Y] = \frac{\partial M(0,0)}{\partial v} = \frac{1}{1-u} \cdot \frac{1}{(1-v)^2}|_{u=v=0} = 1$$

and

$$E[XY] = \frac{\partial^2 M(u,v)}{\partial u \partial v}|_{u=v=0} = \frac{1}{(1-u)^2} \cdot \frac{1}{(1-v)^2}|_{u=v=0} = 1.$$

Thus

$$Cov(X,Y) = E[XY] - E[X]E[Y] = 1 - 1 \times 1 = 0.$$

Example 3.2.18. Let three random variables X, Y and Z are independent and have the moment generating functions

$$M_X(t) = (1-2t)^{-3},$$
$$M_Y(t) = (1-2t)^{-2.5}$$

and

$$M_Z(t) = (1-2t)^{-4.5}$$

respectively. Calculus $E[(X+Y+Z)^3]$.

SOLUTION. Let

$$W = X + Y + Z.$$

We want to find $E[W^3]$. Since X, Y, Z are independent, we have

$$\begin{aligned}
M_W(t) &= E[e^{tW}] = E\left[e^{t(X+Y+Z)}\right] \\
&= E\left[e^{tX}e^{tY}e^{tZ}\right] \\
&= E\left[e^{tX}\right]E\left[e^{tY}\right]E\left[e^{tZ}\right] \\
&= M_X(t)M_Y(t)M_Z(t) \\
&= (1-2t)^{-3}(1-2t)^{-2.5}(1-2t)^{-4.5} \\
&= (1-2t)^{-10}.
\end{aligned}$$

So that

$$\begin{aligned}
M'_W(t) &= (-10)(-2)(1-2t)^{-11} = 20(1-2t)^{-11}, \\
M''_W(t) &= 440(1-2t)^{-12}, \\
M'''_W(t) &= 10,560(1-2t)^{-13}.
\end{aligned}$$

Therefore, by Remark 2.2.7(4),

$$E[W^3] = M'''_W(0) = 10,560$$

CHAPTER 4

Functions of Random Variables

4.1. Distributions of functions of random variables

Quite often we are interested in one or more functions of the random variables associated with some experiment. For example, if $X_1, X_2, \ldots, X_n$ represent repeated measurements of the same random quantity, we might be interested in the maximum and minimum value in the set, as well as the sample mean $\overline{X} = \sum_{i=1}^{n} X_i$ and sample variance $U^2 = \frac{1}{n-1}\sum_{i=1}^{n}(X_i - \overline{X})^2$. In this section we present methods for determining the probabilities of events involving a function $g(X_1, X_2)$ of two random variables X_1, X_2.

First let us find the pdf of a function $Y = u(X)$ of a single random variable X. Suppose that X is a continuous random variable with the distribution function $F_X(x)$ and pdf $f_X(x)$, and suppose that $u(x)$ is a one-to-one function. Then $u(x)$ has an inverse function u^{-1}, so that $u^{-1}(u(x)) = x$. Thus the distribution function $F_Y(y)$ of $Y = u(X)$ is given by

$$F_Y(y) - P(Y \leq y) = P(u(X) \leq y) = P(X \leq u^{-1}(y)) = F_X(u^{-1}(y)).$$

Taking the derivative in both side, we have

$$f_Y(y) = f_X(u^{-1}(y))|(u^{-1})'(y)| = f_X(x)\left|\frac{dx}{dy}\right|. \tag{4.1.1}$$

Similarly, if X is a discrete random variable with the pmf $p_X(x)$, then

$$\begin{aligned} p_Y(y) &= P(Y=y) = P(u(X)=y) = P(X \in u^{-1}(y)) \\ &= \sum_{x_i = u^{-1}(y)} P(X = x_i) = \sum_{x_i = u^{-1}(y)} p_X(x_i). \end{aligned}$$

Example 4.1.1. Let X be a continuous random variable with the pdf

$$f_X(x) = \begin{cases} \frac{1}{x^2}, & x \geq 1 \\ 0, & \text{otherwise.} \end{cases}$$

Find the pdf of the random variable $Y = e^{-X}$.

SOLUTION. Since $y = u(x) = e^{-x}$, we have

$$\frac{dx}{dy} = \frac{1}{dy/dx} = \frac{1}{-e^{-x}} = -e^{x} = -\frac{1}{y}.$$

Thus by the equality (4.1.1),

$$f_Y(y) = f_X(x)\left|\frac{dx}{dy}\right| = \frac{1}{(-\ln y)^2} \cdot \frac{1}{y}, \quad \text{since} \quad x = -\ln y.$$

This function describes the pdf of Y as long as $x \geq 1$, which corresponds to $0 < y \leq e^{-1}$. The complete description of the pdf of Y is

$$f_Y(y) = \begin{cases} \frac{1}{y(\ln y)^2}, & 0 < y \leq e^{-1} \\ 0, & \text{otherwise.} \end{cases}$$

For two random variables X_1, X_2 and a real valued function g defined on $\mathbf{R}^2$, let the random variable Z be defined by $Z = g(X_1, X_2)$. To find the distribution function of Z, note that $\{Z \leq z\} = \{(X_1, X_2) \in A_z\}$, where $A_z = \{(x_1, x_2) \in \mathbf{R}^2 : g(x_1, x_2) \leq z\}$. Then we have

$$\begin{aligned} F_Z(z) &= P(g(X_1, X_2) \leq z) \\ &= P((X_1, X_2) \in A_z) \\ &= \int\int_{A_z} f_{X_1,X_2}(x_1, x_2)\, dx_1 dx_2. \end{aligned}$$

The pdf $f_Z(z)$ of Z is found by taking the derivative of $F_Z(z)$, that is, $f_Z(z) = F_Z'(z)$.

Theorem 4.1.2. *Let X_1 and X_2 be two random variables and $Y = X_1 + X_2$.*
(i) If X_1 and X_2 are discrete non-negative integer valued random variables with the joint pmf $p_{X_1,X_2}(x_1, x_2)$, then the pmf of Y is

$$p_Y(k) = P(Y = k) = \sum_{x_1=0}^{k} p_{X_1,X_2}(x_1, k - x_1).$$

(ii) If X_1 and X_2 are continuous random variables with the joint pdf $f_{X_1,X_2}(x_1, x_2$ then the pdf of Y is

$$f_Y(y) = \int_{-\infty}^{\infty} f_{X_1,X_2}(x_1, y - x_1)dx_1.$$

PROOF. (i) Using the theorem of total probability (1.4.1), we have

$$\begin{aligned} p_Y(k) &= P(X_1 + X_2 = k) \\ &= \sum_{x_1=0}^{k} P(X_1 + X_2 = k | X_1 = x_1) P(X_1 = x_1) \\ &= \sum_{x_1=0}^{k} P(X_1 + X_2 = k, X_1 = x_1) \\ &= \sum_{x_1=0}^{k} P(X_1 = x_1, X_2 = k - x_1) \\ &= \sum_{x_1=0}^{k} p_{X_1,X_2}(x_1, k - x_1). \end{aligned}$$

(ii) Since $\{X_1+X_2 \le y\} = \{(X_1, X_2) \in A_y\}$, where $A_y = \{(x_1, x_2)| -\infty < x_1 < \infty, -\infty < x_2 \le y - x_1\}$, we have

$$\begin{aligned} F_Y(y) &= P(X_1 + X_2 \le y) \\ &= P((X_1, X_2) \in A_y) \\ &= \int\int_{A_y} f_{X_1,X_2}(x_1, x_2) dx_1 dx_2 \\ &= \int_{-\infty}^{\infty} \left\{ \int_{-\infty}^{y-x_1} f_{X_1,X_2}(x_1, x_2) dx_2 \right\} dx_1 \end{aligned}$$

Taking the derivative in both side, we have

$$f_Y(y) = \int_{-\infty}^{\infty} f_{X_1,X_2}(x_1, y - x_1) dx_1.$$

The proof is complete. □

Example 4.1.3. Let X_1 and X_2 be two discrete random variables with the joint pmf $p_{X_1,X_2}(x_1 . x_2)$ which have the following values.

$X_2 \setminus X_1$	0	1	2
4	1/18	3/18	3/18
5	1/18	0	3/18
6	4/18	1/18	2/18

Let $Y = X_1 + X_2$. Find P(Y=6).

SOLUTION. Since the range of X_1 is $\{0, 1, 2\}$, we have by Theorem 4.1.2(i),

$$\begin{aligned} p_Y(6) = P(Y = 6) &= \sum_{x_1=0}^{2} p_{X_1,X_2}(x_1, 6 - x_1) \\ &= p_{X_1,X_2}(0, 6) + p_{X_1,X_2}(1, 5) + p_{X_1,X_2}(2, 4) \\ &= \frac{4}{18} + 0 + \frac{3}{18} \\ &= \frac{7}{18}. \end{aligned}$$

Example 4.1.4. Let X_1 and X_2 be two independent continuous random variables with the same pdf

$$f_{X_1}(x) = f_{X_2}(x) = \begin{cases} \beta e^{-\beta x}, & x > 0 \\ 0, & x < 0. \end{cases}$$

Let $Y = X_1 + X_2$. Show that $Y \sim G(2, \beta)$.

SOLUTION. It is sufficient to show that

$$f_Y(y) = \frac{\beta^2 y e^{-\beta y}}{\Gamma(2)} = \beta^2 y e^{-\beta y}.$$

By Theorem 4.1.2(ii), we see

$$\begin{aligned} f_Y(y) &= \int_{-\infty}^{\infty} f_{X_1,X_2}(x_1, y - x_1) dx_1 \\ &= \int_{-\infty}^{\infty} f_{X_1}(x_1) f_{X_2}(y - x_1) dx_1 \\ &= \int_0^y \beta e^{-\beta x_1} \cdot \beta e^{-\beta(y-x_1)} dx_1 \\ &= \beta^2 \int_0^y e^{-\beta y} dx_1 \\ &= \beta^2 y e^{-\beta y}. \end{aligned}$$

Remark 4.1.5. (1) Let $X_1, X_2, \cdots, X_n$ be random variables and Y a random variable defined to be $Y = \sum_{i=1}^n X_i$. Then, by the same calculus to Remark 3.1.16, we have

$$E[Y] = \sum_{i=1}^n E[X_i]$$

and

$$Var[Y] = \sum_{i=1}^n Var(X_i) + \sum_{i \neq j} Cov(X_i, X_j).$$

If $X_1, X_2, \cdots, X_n$ are independent, then

$$Var[Y] = \sum_{i=1}^n Var(X_i) \quad \text{and} \quad M_Y(t) = \prod_{i=1}^n M_{X_i}(t).$$

(2) If $X_1, X_2, \cdots, X_n$ and $Y_1, Y_2, \cdots, Y_m$ are random variables and $a_1, a_2, \cdots, a_n$, $b_1, b_2, \cdots, b_m$, c and d are constants, then we have

$$Cov\left(\sum_{i=1}^{n} a_i X_i + c, \sum_{j=1}^{m} b_j Y_j + d\right) = \sum_{i=1}^{n}\sum_{j=1}^{m} a_i b_j Cov(X_i, Y_j).$$

Example 4.1.6. ([2]) Let $X_1, X_2, \cdots, X_n$ be n independent random variables with the same variance σ^2, and let $U = 2X_1 + X_2 + \cdots + X_{n-1}$ and $V = X_2 + \cdots + X_{n-1} + 2X_n$. Find the coefficient of correlation between U and V.

SOLUTION. We have

$$\begin{aligned} Var(U) &= 4Var(X_1) + Var(X_2) + \cdots + Var(X_{n-1}) \\ &= (n+2)\sigma^2. \end{aligned}$$

And similarly,

$$\begin{aligned} Var(V) &= Var(X_2) + \cdots + Var(X_{n-1}) + 4Var(X_{n-1}) \\ &= (n+2)\sigma^2. \end{aligned}$$

Since $X_1, X_2, \cdots, X_n$ are independent, $Cov(X_i, X_j) = 0$ if $i \neq j$. Noting that $Cov(X, X) = Var(X)$, we have

$$\begin{aligned} Cov(U, V) &= Cov(2X_1, X_2) + \cdots + Cov(X_{n-1}, 2X_n) \\ &= Var(X_2) + \cdots + Var(X_{n-1}) \\ &= (n-2)\sigma^2. \end{aligned}$$

Thus

$$\begin{aligned} \rho_{U,V} &= \frac{Cov(U,V)}{\sqrt{Var(U)}\sqrt{Var(V)}} \\ &= \frac{(n-2)\sigma^2}{(n+2)\sigma^2} \\ &= \frac{n-2}{n+2}. \end{aligned}$$

Next result is important in actuarial sciences.

Theorem 4.1.7. *Let X_i, $i = 1, 2, \cdots$ is a sequence of independent random variables having the same distribution with a random variable X, and let*

$$Y = X_1 + X_2 + \cdots + X_N,$$

where N is assumed to be a random variable that is independent of the X_j's. Then we have

(i) $E[Y] = E[N]E[X]$.

(ii) $Var(Y) = Var(N)E[X]^2 + E[N]Var(X)$.

PROOF. (i) First let us find the mgf $M_Y(t)$ of Y.

$$\begin{aligned} E\left[e^{tY}|N=n\right] &= E\left[\exp\left\{t\sum_{i=1}^{N} X_i\right\}|N=n\right] \\ &= E\left[\exp\left\{t\sum_{i=1}^{n} X_i\right\}\right] \\ &= E\left[\Pi_{i=1}^{n} e^{tX_i}\right] \\ &= \Pi_{i=1}^{n} E\left[e^{tX_i}\right] \\ &= E\left[e^{tX}\right]^n \\ &= M_X(t)^n. \end{aligned}$$

Hence we have

$$E\left[e^{tY}|N\right] = M_X(t)^N.$$

By properties of the conditional expectation, we have

$$\begin{aligned} M_Y(t) &= E\left[e^{tY}\right] \\ &= E\left[E\left[e^{tY}|N\right]\right] \\ &= E\left[M_X(t)^N\right]. \end{aligned}$$

Taking the derivative in both side,

$$M_Y'(t) = E\left[N M_X^{N-1}(t) M_X'(t)\right] \tag{4.1.2}$$

From this we have

$$\begin{aligned} E[Y] &= M_Y'(0) \\ &= E\left[N M_X(0)^{N-1} M_X'(0)\right]. \end{aligned}$$

Since $M_X(0) = 1$ and $M_X'(0) = E[X]$,

$$\begin{aligned} E[Y] &= E[NE[X]] \\ &= E[N]E[X]. \end{aligned}$$

Now taking the derivative in both side of (4.1.2),

$$M_Y''(t) = E\left[N(N-1)M_X(t)^{N-2}M_X'(t)^2 + NM_X(t)^{N-1}M_X''(t)\right],$$

so that

$$\begin{aligned} E\left[Y^2\right] &= M_Y''(0) \\ &= E\left[N(N-1)M_X(0)^{N-2}{M_X'}^2(0) + NM_X(0)^{N-1}M_X''(0)\right] \\ &= E\left[N(N-1)E[X]^2 + NE[X^2]\right] \\ &= E\left[N\left(E[X^2] - E[X]^2\right) + N^2E[X]^2\right] \\ &= E[N]Var(X) + E[N^2]E[X]^2. \end{aligned}$$

These imply

$$\begin{aligned} Var(Y) &= E[Y^2] - E[Y]^2 \\ &= E[N]Var(X) + E[N^2]E[X]^2 - E[N]^2E[X]^2 \\ &= E[N]Var(X) + (E[N^2] - E[N]^2)E[X]^2 \\ &= E[N]Var(X) + Var(N)E[X]^2. \end{aligned}$$

The proof is complete. □

4.2. The central limit theorem

Let $X_1, X_2, \cdots$ be a sequence of independent random variables with same mean μ and same variance σ^2, and let Y_n be the sum of the first n

random variables in the sequence:

$$Y_n = X_1 + X_2 + \cdots + X_n.$$

In this section, we present the central limit theorem, which state, as n becomes large, the distribution function of Y_n approaches that of a normal random variable. The central limit theorem explains why the normal random variable appear in so many diverse applications.

From Remark 4.1.5(1), we know that if X_j's are **iid random variables**, i.e., they are independent and have identical distributions, then Y_n has mean $n\mu$ and variance $n\sigma^2$. The central limit theorem says that the pdf of the normalized version of Y_n approaches that of a standard normal random variable.

Theorem 4.2.1. *Let $X_1, X_2, \cdots$ be a sequence of iid random variables with same mean μ and same variance σ^2, and let*

$$Y_n = X_1 + X_2 + \cdots + X_n.$$

Then
(i) $E[Y_n] = n\mu$ and $Var(Y_n) = n\sigma^2$.
(ii) If

$$Z_n = \frac{Y_n - n\mu}{\sqrt{n}\sigma},$$

then $Z_n \sim N(0, 1^2)$ approximately in the sense that

$$\lim_{n\to 0} P(Z_n \leq z) = \Phi(z) = \int_0^z \frac{1}{2\pi} e^{-x^2/2} dx.$$

(iii) For any sufficient large n,

$$Y_n \sim N(n\mu, n\sigma^2) \quad \textit{approximately.}$$

PROOF. The result (i) is clear by Remark 4.1.5(1).

Let us prove (ii). Let $M(t)$ be the common mgf of the X_i's. Since $X_1, X_2 \cdots, X_n$ are independent, we have

$$\begin{aligned} M_{Y_n}(t) &= E\left[e^{tY_n}\right] \\ &= E\left[e^{t\Sigma_{i=1}^n X_n}\right] \\ &= E\left[e^{tX_1} \cdots e^{tX_n}\right] \\ &= E\left[e^{tX_1}\right] \cdots E\left[e^{tX_n}\right] \\ &= M(t)^n. \end{aligned}$$

So we have

$$\begin{aligned} M_{Z_n}(t) &= E\left[e^{tZ_n}\right] \\ &= E\left[e^{\frac{t(-n\mu+Y_n)}{\sqrt{n}\sigma}}\right] \\ &= e^{\frac{-\sqrt{n}t\mu}{\sigma}} E\left[e^{t\frac{Y_n}{\sqrt{n}\sigma}}\right] \\ &= e^{\frac{-\sqrt{n}t\mu}{\sigma}} M_{Y_n}\left(\frac{t}{\sqrt{n}\sigma}\right) \\ &= e^{\frac{-\sqrt{n}t\mu}{\sigma}} \left(M\left(\frac{t}{\sqrt{n}\sigma}\right)\right)^n. \end{aligned}$$

Taking ln in both side,

$$lnM_{Z_n}(t) = -\frac{\sqrt{n}\mu}{\sigma}t + n\ln M\left(\frac{t}{\sqrt{n}\sigma}\right).$$

Expanding $M(t)$ in the Maclaurin series, it holds

$$M(t) = 1 + M'(0)t + \frac{M''(0)}{2!}t^2 + R,$$

where R is the remainder term. Recalling that $M'(0) = E[X_1] = \mu$ and $M''(0) = E[X_1^2] = E[X_1]^2 + Var(X_1) = \mu^2 + \sigma^2$, we obtain

$$M_X(t) = 1 + \mu t + \frac{(\mu^2+\sigma^2)}{2}t^2 + R.$$

Hence

$$\ln M_{Z_n}(t) = -\frac{\sqrt{n}\mu}{\sigma}t + n\ln\left[1 + \frac{\mu t}{\sqrt{n}\sigma} + \frac{(\mu^2+\sigma^2)t^2}{2n\sigma^2} + R\right].$$

We shall now use the Maclaurin expansion for $\ln(1+x)$:

$$\ln(1+x) = x - \frac{x^2}{2} + \frac{x^3}{3} - \cdots, \quad |x| < 1.$$

In our case,

$$x = \frac{\mu t}{\sqrt{n}\sigma} + \frac{(\mu^2+\sigma^2)t^2}{2n\sigma^2} + R$$

and for n sufficiently large, the absolute value of this expression will be less than one. Hence we have

$$\ln M_{Z_n}(t) = -\frac{\sqrt{n}\mu t}{\sigma} + n\left[\left\{\frac{\mu t}{\sqrt{n}\sigma} + (\mu^2+\sigma^2)\frac{t^2}{2n\sigma^2} + R\right\} \right.$$
$$\left. -\frac{1}{2}\left\{\frac{\mu t}{\sqrt{n}\sigma} + (\mu^2+\sigma^2)\frac{t^2}{2n\sigma^2} + R\right\}^2 + \cdots\right].$$

We can see that the right hand side converges to $t^2/2$ as $n \to \infty$, that is,

$$\lim_{n\to\infty} \ln M_{Z_n}(t) = \frac{t^2}{2}.$$

Thus we have

$$\lim_{n\to\infty} M_{Z_n}(t) = e^{\frac{t^2}{2}}.$$

This is the mgf of a standard random variable. Because of the uniqueness property of the mgf we may conclude that the distribution function of Z_n converges to the distribution function of a standard random variable as $n \to \infty$.

Finally we prove (iii). We just have proved that $Z_n \sim N(0, 1^2)$ for sufficiently large n. Since $Y_n = n\sqrt{\sigma}Z_n + n\mu$, by Theorem 2.4.5, we see that $Y_n \sim N(n\mu, n\sigma^2)$ for sufficiently large n. The proof is complete. □

Corollary 4.2.2 (The central limit theorem). *For given n iid random variables $X_1, X_2, \cdots, X_n$ with same mean μ and same variance σ^2, define*

$$\bar{X} = \frac{1}{n}\sum_{i-1}^{n} X_i.$$

Then

$$\frac{\bar{X}-\mu}{\sigma/\sqrt{n}} \sim N(0, 1^2) \quad \textit{approximately.} \tag{4.2.1}$$

and

$$\bar{X} \sim N\left(\mu, \frac{\sigma^2}{n}\right) \quad \textit{approximately.} \tag{4.2.2}$$

PROOF. By a simple calculation, we obtain

$$\frac{\bar{X} - \mu}{\sigma/\sqrt{n}} = \frac{\frac{1}{n}\Sigma_{i=1}^n X_i - \mu}{\sigma/\sqrt{n}} = \frac{\Sigma_{i=1}^n X_i - n\mu}{\sqrt{n}\sigma}.$$

Since right hand side equals Z_n in Theorem 4.2.1, (4.2.1) is proved. From (4.2.1) and the same argument as Theorem 4.2.1(iii), we can see that (4.2.2) is true. The proof is complete. □

Example 4.2.3. A fair die rolled 420 times. What is the probability that the sum of the rolls lies between 1400 and 1540?

SOLUTION. The sum is random variable

$$Y_{420} = X_1 + X_2 + \cdots + X_{420},$$

where each X_j has the same pmf $p_{X_j}(k) = 1/6$ for all $k = 1, 2, \cdots, 6$. Hence, for each $j = 1, 2, \cdots, 420$,

$$\begin{aligned} E[X_j] &= 1 \times \frac{1}{6} + 2 \times \frac{1}{6} + \cdots + 6 \times \frac{1}{6} \\ &= \frac{7}{2} (= \mu) \end{aligned}$$

and

$$\begin{aligned} Var(X_j) &= E[X_j^2] - E[X_j]^2 \\ &= 1^2 \times \frac{1}{6} + 2^2 \times \frac{1}{6} + \cdots + 6^2 \times \frac{1}{6} - \left(\frac{7}{2}\right)^2 \\ &= \frac{35}{12} (= \sigma^2) \end{aligned}$$

Thus

$$E[Y_{420}] = 420 \cdot \frac{7}{2} = 1470 (= n\mu)$$

and

$$Var(Y_{420}) = 420 \cdot \frac{35}{12} = 1225 = 35^2 (= n\sigma^2)$$

By Theorem 4.2.1(iii), $Y_{420} \sim N(1470, 35^2)$ approximately. Therefore,

$$\begin{aligned}
&P(1400 \leq Y_{420} \leq 1540) \\
&= P\left(\frac{1400-1470}{35} \leq \frac{Y_{420}-1470}{35} \leq \frac{1540-1470}{35}\right) \\
&= P\left(-2 \leq \frac{Y_{420}-1470}{35} \leq 2\right) \\
&= \Phi(2) - \Phi(-2) \\
&= 2\Phi(2) - 1 \\
&= 2 \cdot 0.9772 - 1 \\
&= 0.9544.
\end{aligned}$$

Example 4.2.4. Suppose that $X \sim B(1, 0.5)$ and $X, X_1, X_2, \cdots, X_{25}$ are iid random variables. Let

$$Y_{25} = X_1 + X_2 + \cdots + X_{25}.$$

Estimate $P(Y_{25} \leq 12)$.

SOLUTION. Since $E[X] = 1 \times 0.5 = 0.5$ and $Var(X) = 1 \times 0.5 \times 0.5 = (0.5)^2$, we have $E[Y_{25}] = 25 \times 0.5 = 12.5$ and $Var(Y_{25}) = 25 \times (0.5)^2 = 5^2 \times (0.5)^2 = (2.5)^2$. Thus

$$\begin{aligned}
P(Y_{25} \leq 12) &= P\left(\frac{Y_{25}-12.5}{2.5} \leq \frac{12-12.5}{2.5}\right) \\
&= P\left(\frac{Y_{25}-12.5}{2.5} \leq -0.2\right) \\
&\approx \Phi(-0.2) \\
&= 1 - \Phi(0.2) \\
&= 1 - 0.5793 \\
&= 0.4207.
\end{aligned}$$

But since Y_{25} is an integer valued random variable, using the continuous correction (see Remark 2.4.12), we have

$$\begin{aligned} P(Y_{25} \leq 12) &= P(Y_{25} < 12.5) \\ &= P\left(\frac{Y_{25} - 12.5}{2.5} \leq \frac{12.5 - 12.5}{2.5}\right) \\ &\approx \Phi(0) = 0.5, \end{aligned}$$

which is the exact probability of at most 12 successes in 25 trials with a fair coin.

Example 4.2.5. Suppose that $X \sim P(1)$ and $X, X_1, X_2, \cdots, X_{64}$ are iid random variables. If $Y_{64} = X_1 + X_2 + \cdots + X_{64}$, find $P(50 < Y_{64} \leq 80)$.

SOLUTION. Since Y_{64} is an integer valued random variable, by the continuous correction again, we have

$$\begin{aligned} P(50 < Y_{64} \leq 80) &= P(51 \leq Y_{64} \leq 80) \\ &= P(50.5 < Y_{64} < 80.5). \end{aligned}$$

Since $E[X] = Var(X) = 1$, we have $E[Y_{64}] = 64E[X] = 64$ and $Var(Y_{64}) = 64Var(X) = 64 = 8^2$. Thus

$$\begin{aligned} P(50 < Y_{64} \leq 80) &= P(50.5 < Y_{64} < 80.5) \\ &= P(\frac{50.5 - 64}{8} < \frac{Y_{64} - 64}{8} < \frac{80.5 - 64}{8}) \\ &= P(-1.69 < \frac{Y_{64} - 64}{8} < 2.06) \\ &\approx \Phi(2.06) - \Phi(-1.69) \\ &= \Phi(2.06) + \Phi(1.69) - 1 \\ &= 0.9803 + 0.9545 - 1 \\ &= 0.9348. \end{aligned}$$

CHAPTER 5

Applications to Loss Distributions and Insurance

5.1. How insurance works?

Modern society provides many examples of **risk** of incurring a financial loss. For example,

- A homeowner faces a large potential for variation associated with the possibility of economic loss caused by a house fire.

- A driver faces a potential economic loss if his car is damaged.

- A family faces the risk of incurring a financial loss by dental expenses for family.

Someone who is at risk of financial loss may choose some form of **insurance** protection to reduce the impact of the loss. An **insurance policy** is a contract between the party that is at risk(= **policyholder** or **insured**) and the insurance company(= **insurer**). The insurance policy calls for the policyholder to pay the insurer some specified amount(= **insurance premium**), and in return, the insurer will reimburse a defined amount(= **claim** or **benefit**).

The loss is generally modeled using a random variable or some combination of random variables. *Unless indicated otherwise, it will be assumed that the claim amount is equal to the the amount of the loss that occurs.*

When the loss or claim random variable is denoted by X, the **expected claim** $E[X]$ on the insurer is referred to as the **pure** or **net premium** for the policy.

For a loss random variable X, a measure of the risk is $\sqrt{Var(X)}$. In this case, $\sqrt{Var(X)}/E[X]$ is called the **unitized risk** or **coefficient of variation** for X.

5.2. Models for describing a loss random variable

There are three cases of models for describing a loss random variable X:
(I) The case that the complete description of X is given.
(II) The case that the probability q that the loss occurs is given, and the conditional distribution of a loss amount X given that a loss has occurred, say the distribution of B, is given.
(III) The case that the number N of losses on a particular policyholder is a random variable, and the claim amount B per loss is a random variable.
We will show that we can find $E[X]$, $Var(X)$ and hence the premium in each case.

The case (I).

Consider the case that the complete description of X is given. If X is continuous random variable, then the distribution function $F_X(x)$ of the pdf $f_X(x)$ is given. If X is discrete random variable, the pmf $p_X(x)$ is given. In some problems, $E[X]$ and $Var(X)$ are needed, and sometimes only these, rather than the full description of the distribution of X, are the information about X that we can use.

Example 5.2.1. Consider a one-year term life insurance paying an extra benefit in case of accidental death. To be specific, if death is accidental, the benefit amount is 50,000. For other causes of death, the benefit amount is 25,000. Let us assume that for the age, health and occupation of specific individual the probability of an accidental death within the year is 0.0005 while the probability of a nonaccidental death is 0.002.

Find $E[X]$ and $Var(X)$ of the claim benefit X.

SOLUTION. The complete description of the claim benefit X is given as follows;

$$X = \begin{cases} 50,000 & \text{with probability } \; 0.0005 \\ 25,000 & \text{with probability } \; 0.002 \\ 0 & \text{with probability } \; 0.9975 \end{cases}$$

By the definition of the expectation and variance of X, we have

$$E[X] = (50,000)(0.0005) + (25,000)(0.002) + (0)(0.9975) = 75$$

and

$$\begin{aligned} Var(X) &= E[X^2] + E[X]^2 \\ &= (50,000)^{(}0.0005) + (25,000)^{(}0.002) + (0)^{(}0.9975) - 75^2 \\ &= 2,500,000 - 5625 = 2,494,375. \end{aligned}$$

The case (II).

Consider the case that the probability q that the loss occurs and the distribution $F_B(x)$ or pdf $f_B(x)$ of B, instead of the claim amount X, such that $F_B(x) = F_{X|I}(x|I=1)$ or $f_B(x) - f_{X|I}(x|I=1)$ are given, where

$$I = \begin{cases} 1, & \text{if loss occurs, with probability } \; q \\ 0, & \text{if loss does not occur, with probability } \; 1-q. \end{cases}$$

In this case we have

$$\begin{aligned} &E[X] = qE[B] \\ (5.2.1) \quad & \\ &Var(X) = qE[B^2] - (qE[B])^2 = q(1-q)E[B]^2 + qVar(B). \end{aligned}$$

To prove these we use the properties of the conditional expectation. Since

$$E[X|I] = \begin{cases} E[X|I=1], & \text{with probability} \quad P(I=1) \\ E[X|I=0], & \text{with probability} \quad P(I=0), \end{cases}$$

$$E[E[X|I]] = E[X|I=1]P(I=1) + E[X|I=0]P(I=0) \tag{5.2.2}$$

The first term of the right hand side is given by

$$\begin{aligned} &q\int_{-\infty}^{\infty} x f_{X|I}(x|I=1)dx \\ &= q\int_{-\infty}^{\infty} x f_B(x)dx \\ &= qE[B]. \end{aligned}$$

If a loss does not occur, then the claim amount is 0, i.e., $X = 0$. This shows the second term of the right hand side of (5.2.2) is 0. So we have

$$E[X] = E[E[X|I]] = qE[B].$$

Now we prove the second equality of (5.2.1). By the same calculus as the proof of the first equality, it holds

$$\begin{aligned} E[E[X^2|I]] &= E[X^2|I=1]P(I=1) + E[X^2|I=0]P(I=0) \\ &= q\int_{-\infty}^{\infty} x^2 f_{X|I}(x|I=1)dx + 0 \\ &= q\int_{-\infty}^{\infty} x^2 f_B(x)dx \\ &= qE[B^2]. \end{aligned}$$

Thus we have

$$E[X^2] = E[E[X^2|I]] = qE[B^2]$$

and, hence

$$Var(X) = E[X^2] - E[X]^2 = qE[B^2] - q^2E[B]^2.$$

Example 5.2.2. Let X denote the claim benefit described in Example 5.2.1. Use (5.2.1) to find $E[X]$ and $Var(X)$.

SOLUTION. Let q be the probability that a loss occurs, i.e., the probability of death, and let B denote a claim amount given that a death has occurred. Then we have

$$\begin{aligned} q &= P(I=1) \\ &= P(I=1, X=50,000) + P(I=1, X=25,000) \\ &= 0.0005 + 0.002 = 0.0025. \end{aligned}$$

Since the pmf $p_B(x)$ of the random variable B is given by

$$\begin{aligned} p_B(x) &= p_{X|I}(x|I=1) = P(X=x|I=1) \\ &= \frac{P(I=1, X=x)}{P(I=1)} \end{aligned}$$

for $x = 50,000$ and 25,000, we get

$$p_B(50,000) = \frac{P(I=1, X=50,000)}{P(I=1)} = \frac{0.0005}{0.0025} = 0.2$$

and

$$p_B(25,000) = \frac{P(I=1, X=25,000)}{P(I=1)} = \frac{0.002}{0.0025} = 0.8$$

Thus the random variable B is described by

$$B = \begin{cases} 50,000 & \text{with probability } 0.2 \\ 25,000 & \text{with probability } 0.8 \end{cases}$$

From this we have

$$E[B] = (50,000)(0.2) + (25,000)(0.8) = 30,000$$

and

$$Var(B) = (50,000)^2(0.2) + (25,000)^2(0.8) - (30,000)^2 = 10,000^2.$$

Using (5.2.1), we have

$$E[X] = qE[B] = (0.0025)(30,000) = 75$$

and

$$\begin{aligned}
Var(X) &= q(1-q)E[B]^2 + qVar(B) \\
&= (0.0025)(0.9975)(30,000)^2 + (0.0025)(10,000)^2 \\
&= 2244375 + 250,000 = 2,494,375.
\end{aligned}$$

Example 5.2.3. ([2]) Suppose that $q = 0.01$ is the probability that fire damage occurs, and given that fire damage occurs, the amount of damage B has a uniform distribution between \$10,000 and \$50,000. For the loss amount X, find $E[X]$ and $Var(X)$. (Keep in mind that B is the loss amount given that a loss has occurred, where X is the unconditional loss amount.)

SOLUTION. Since $B \sim U(10,000,\ 50,000)$, we have

$$E[B] = \frac{10,000 + 50,000}{2} = 30,000$$

and

$$E[B^2] = \frac{((10,000)^2 + (10,000)(50,000) + (50,000)^2}{3} = \frac{3,100,000,000}{3}.$$

Using (5.2.1), we have

$$E[X] = qE[B] = (0.01)(30,000) = 300$$

and

$$\begin{aligned}
Var(X) &= qE[B^2] + (qE[B])^2 \\
&= (0.01)\left(\frac{3,100,000,000}{3}\right) + (0.01)^2(30,000)^2 \\
&= \frac{30,730,000}{3}.
\end{aligned}$$

The case (III).

Consider the case that the number of losses that occur on a particular policyholder is specified as a nonnegative integer valued random number N and the loss amount per loss is specified as a random variable B. In this case the total loss amount X is given by $X = 0$ if $N = 0$, and

$$X = B_1 + B_2 + \cdots + B_N,$$

where B_k's are independent and identically distributed(**iid**) random variables with B. Under these conditions, we have

$$\begin{aligned} E[X] &= E[N]E[B] \\ Var(X) &= Var(N)E[B]^2 + E[N]Var(B). \end{aligned} \tag{5.2.3}$$

Note that the case (II) is a special form of the case (III), in which N takes only two values 0 and 1 and so $E[N] = 0 \cdot (1-q) + 1 \cdot q = q$.

Example 5.2.4. ([**2**]) For a particular one year auto insurance policy, assume that the number N of auto accidents occurring in one year has a geometric distribution with parameter $p = 0.9$, and assume that if an accident occur, then the damage claim amount B has a uniform distribution on the interval $(0, 10000)$.

If X represents the total amount claimed on the particular policy, find $E[X]$ and $Var(X)$.

Solution. Since $N \sim G(0.9)$, i.e., $p_N(k) = (1-0.9)^k(0.9)$, we get

$$E[N] = \frac{1-0.9}{0.9} = \frac{1}{9}$$

and

$$Var(N) = \frac{1-0.9}{0.9^2} = \frac{10}{81}$$

And since $B \sim U(0,\ 10,000)$, we have

$$E[B] = \frac{0+10,000}{2} = 5,000$$

and

$$Var(B) = \frac{(10,000-0)^2}{12} = 8,333,333.$$

Using (5.2.3), we have

$$E[X] = E[N]E[B] = \frac{1}{9} \times 5,000 = 556$$

and

$$\begin{aligned} Var(X) &= Var(N)E[B]^2 + E[N]Var(B) \\ &= \frac{10}{81} \times 5,000^2 + \frac{1}{9} \times 8,333,333 \\ &= 4,012,346. \end{aligned}$$

5.3. Modeling the aggregate claims

There are two ways of modeling the aggregate claims for a period on a portfolio of insurance policies. They are the individual risk model and the collective risk model.

The **individual risk model** assumes that the portfolio consists of a specific number, say n, of insurance policies, with the claim for one period on policy k being the random variable X_k that is modeled in one of the ways described in Section 5.2. Unless mentioned otherwise, it is assumed that the X_k's are independent. Then the aggregate claim is a random variable

$$S = X_1 + X_2 + \cdots + X_n.$$

If $E[X_k] = \mu$ and $Var(X_k) = \sigma^2$ for each $k = 1, 2, \cdots, n$, then

$$E[S] = \sum_{k=1}^{n} E[X_k] = n\mu \quad \text{and} \quad Var(S) = \sum_{k=1}^{n} Var(X_k) = n\sigma^2.$$

Example 5.3.1. ([**2**]) An insurer has a portfolio of 1000 one-year term life insurance policies just issued to 1000 independent individuals. Each policy will be paid \$1000 in the event that the policyholder dies within the year. For 500 of the policies, the probability of death is 0.01 per policyholder, and for the other 500 policies the probability of death is 0.02 per policyholder. Find the expectation and the variance of the aggregate claim that the insurer will pay.

SOLUTION. Let S be the aggregate claim that the insurer will pay, and let Class I be the 500 policies with death probability 0.01 and Class II another 500 policies with death probability 0.02. Then since

$$S = \sum_{i=1}^{1,000} X_i = \sum_{\text{Class I}}^{500} X_i + \sum_{\text{Class II}}^{500} X_i,$$

$$E[S] = \sum_{\text{Class I}}^{500} E[X_i] + \sum_{\text{Class II}}^{500} E[X_i]$$

and

$$Var(S) = \sum_{\text{Class I}}^{500} Var(X_i) + \sum_{\text{Class II}}^{500} Var(X_i).$$

For any $i = 1, 2, \cdots$, if X_i is a claim by i-th policy in Class I, it is described by

$$X_i = \begin{cases} 0, & \text{w.p. } 0.99 \\ 1,000, & \text{w.p. } 0.01. \end{cases}$$

Thus

$$E[X_i] = (0)(0.99) + (1,000)(0.01) = 10$$

and

$$Var(X_i) = (0)^2(0.99) + (1,000)^2(0.01) - 10^2 = 9,900.$$

By the same calculation, for any $i = 1, 2, \cdots$, if X_i is a claim by i-th policy in Class II, then $E[X_i] = 20$ and $Var(X_i) = 19,600$. Therefore we have

$$E[S] = \sum_{\text{Class I}}^{500} E[X_i] + \sum_{\text{Class II}}^{500} E[X_i] = \sum_{\text{Class I}}^{500} 10 + \sum_{\text{Class II}}^{500} 20 = 500 \times 10 + 500 \times 20 = 15,000$$

and

$$\begin{aligned} Var(S) &= \sum_{\text{Class I}}^{500} Var(X_i) + \sum_{\text{Class II}}^{500} Var(X_i) \\ &= \sum_{\text{Class I}}^{500} 9,900 + \sum_{\text{Class II}}^{500} 19,600 \\ &= 500 \times 9,900 + 500 \times 19,600 = 14,750,000. \end{aligned}$$

Alternatively, we can get $E[X_i]$ and $Var(X_i)$ by another method. For any i-th policy in Class I, since $q_i = 0.01$ and $B_i = 1,000$, we have

$$E[X_i] = q_i E[B_i] = 0.01 \times 1,000 = 10$$

and

$$\begin{aligned} Var(X_i) &= q_i E[B_i^2] - (q_i E[B_i])^2 \\ &= 0.01 \times 1,000^2 - (0.01 \times 1,000)^2 = 9,900. \end{aligned}$$

By the same calculation, for any i-th policy in Class II, we have $E[X_i] = 20$ and $Var(X_i) = 19,600$.

The **collective risk model** specifies a nonnegative integer valued random variable N to describe the number(or frequency) of positive claims that occur out of the portfolio of policies in one period. In this model, the aggregate claim S is given by $S = 0$ if $N = 0$, and

$$S = Y_1 + Y_2 + \cdots + Y_N \quad \text{if} \quad N > 0,$$

where Y_i is the amount of i-th claim, and Y_i's are iid with a random variable Y. Then by Theorem 4.1.7, we have

$$\begin{aligned} E[S] &= E[N]E[Y] \\ Var(S) &= Var(N)E[Y]^2 + E[N]Var(Y). \end{aligned} \tag{5.3.1}$$

The distribution of Y is called the **severity distribution**.

Example 5.3.2. Suppose that the number N of claims per period has a geometric distribution with parameter $p = 0.9$ and that the claim amount

Y has a uniform distribution on the interval $(0, 10000)$. If S is the aggregate claim, find $E[S]$ and $Var(S)$.

SOLUTION. This situation becomes mathematically identical to the situation in Example 5.2.4, where N was the number of claims that the individual had, and B, which has the same distribution as Y, was the claim amount given that an accident occurs. Therefore $E[S] = 556$ and $Var(S) = 4,012,346$.

Since the aggregate claim random variable S is a sum of many independent individual policy loss random variables, by the Central Limit Theorem, we can consider that $S \sim N(E[S], Var(S))$ approximately. Hence, for example, it is possible to find the number Q such that $P(S \leq Q) = 0.95$. In fact,

$$0.95 = P(S \leq Q) = P\left(\frac{S - E[S]}{\sqrt{Var(S)}} \leq \frac{Q - E[S]}{\sqrt{Var(S)}}\right).$$

From the standard normal table, we have $(Q - E[S])/\sqrt{Var(S)} = 1.645$ and so $Q = E[S] + 1.645\sqrt{Var(S)}$. The amount Q is referred to the **total premium** that the insurer collects.

For the aggregate claim S, the total premium Q can be sometimes written in the form

$$Q = (1 + \theta)E[S].$$

The factor θ is called the **relative security loading** contained within the premium. Thus we can say that the premium consists the expected aggregate claim $E[S]$ plus a loading $\theta \cdot E[S]$ on top of the expected aggregate claim.

Example 5.3.3. ([**1**]) A life insurance company issues 1-year term life contracts for benefit amounts of 1 and 2 units to individuals with probabilities of death of 0.02 or 0.10. The following table gives the number of individuals n_k in each of the four classes created by a benefit amount B_k and a probability of claim q_k.

Class k	Number in Class (n_k)	Probability of claim per policy (q_k)	Claim amount (B_k)
1	500	0.02	1
2	500	0.02	2
3	300	0.10	1
4	500	0.10	2

The company wants to collect, from this population of 1,800 individuals, an amount equal to the 95th percentile of the distribution of total claims.

Calculate the relative security loading θ.

SOLUTION. The criterion for θ is $P(S \leq (1+\theta)E[S]) = 0.95$ where $S = X_1 + X_2 + \cdots + X_{1800}$. This probability statement is equivalent to

$$P\left(\frac{S - E[S]}{\sqrt{Var(S)}} \leq \frac{\theta E[S]}{\sqrt{Var(S)}}\right) = 0.95$$

Following the discussion of the central limit theorem, we obtain

$$\frac{\theta E[S]}{\sqrt{Var(S)}} = 1.645.$$

It remains to calculate the mean and variance of S and to calculate θ by this equation. For any j-th policy in Class k, since $B_j = b_k =$ constant 1 or 2 , we have $E[B_j] = b_k$ and $Var(B_j) = 0$. Thus for any j-th policy in Class k, it holds

$$E[X_j] = q_j E[B_j] = q_k b_k$$

and

$$Var(X_j) = g_j(1-q_j)E[B_j]^2 + q_j Var(B_j) = g_k(1-q_k)b_k^2.$$

For the four classes of insured individuals, we have the results given below.

k	n_k	q_k	$B_k = b_k$	$E[X_k]$	$Var(X_k)$
1	500	0.02	1	0.02	0.0196
2	500	0.02	2	0.04	0.0784
3	300	0.10	1	0.10	0.0900
4	500	0.10	2	0.20	0.3600

Then

$$E[S] = \sum_{j=1}^{1800} E[X_j] = \sum_{k=1}^{4} n_k E[X_k] = 160$$

and

$$Var(S) = \sum_{j=1}^{1800} Var(X_j) = \sum_{k=1}^{4} n_k Var(X_k) = 256 = 16^2.$$

Thus, the relative security loading is

$$\theta = 1.645 \frac{\sqrt{Var(S)}}{E[S]} = 1.645 \frac{16}{160} = 0.1645.$$

Example 5.3.4. ([**1**]) The policyholders of an automobile insurance company fall into two classes.

Class k	Number in Class n_k	Claim probability q_k	Distribution of claim amount if a loss occurs $B_k \sim TE(\lambda_k, L_k)$ λ_k	L_k
1	500	0.10	1	2.5
2	2,000	0.05	2	5.0

The probability that total claims exceed the amount collected from policyholders is to be 0.05. We assume that the relative security loading, θ, is to be same for the two classes. Calculate θ.

SOLUTION. This example is much like the previous one. It differs in that the claim amounts are random variables. By Remark 2.4.22, we see that if $B \sim TE(\lambda, L)$, then

$$E[B] = \frac{1 - e^{-\lambda L}}{\lambda}$$

and

$$Var(B) = \frac{1 - 2\lambda L e^{-\lambda L} - e^{-2\lambda L}}{\lambda^2}.$$

Using the parameter values given and applying formulas in (5.2.1), we obtain the following results.

k	n_k	q_k	$E[B_k]$	$Var(B_k)$	$E[X_k]$	$Var(X_k)$
1	500	0.10	0.9179	0.5828	0.09179	0.13411
2	2,000	0.05	0.2498	0.2498	0.02500	0.02436

Then S, the sum of the claim, has moments

$$\begin{aligned} E[S] &= \sum_{j=1}^{2500} E[X_j] = \sum_{k=1}^{2} n_k E[X_k] \\ &= 500(0.09179) + 2,000(0.02500) = 95.89 \end{aligned}$$

and

$$\begin{aligned} Var(S) &= \sum_{j=1}^{2500} Var(X_j) = \sum_{k=1}^{2} n_k Var(X_k) \\ &= 500(0.13411) + 2,000(0.02436) = 115.78. \end{aligned}$$

The criterion for θ is the same as in Example 5.3.3,

$$P(S \leq (1+\theta)E[S]) = 1 - 0.05 = 0.95.$$

Again by the normal approximation,

$$\frac{\theta E[S]}{\sqrt{Var(S)}} = 1.645$$

and

$$\theta = \frac{1.645\sqrt{115.78}}{95.89} = 0.1846.$$

5.4. The effect of coverage adjustments on a loss distribution

It is possible to construct an insurance policy in which *the claim paid by the insurer is part, but not necessarily all, of the loss that occurs.* Let X denote the loss amount that would be paid by an insurance if the loss were fully covered. X is known as the **ground-up loss**. There are standard types of partial insurance coverage on X such as an excess-of-loss insurance, a policy limit and a proportional insurance.

An **excess-of-loss insurance** specifies a **deductible amount** d. If a loss amount X occurs, the amount paid by the insurer can be described as

$$Y_l = \begin{cases} 0, & \text{if} \quad X \le d \\ X - d, & \text{if} \quad X > d. \end{cases}$$

In this case Y_l is the **amount paid per loss** and $Y_c = (X - d)|_{X>d}$ is the **amount paid per claim** (i.e., per payment).

Remark 5.4.1. Clearly we can see that a loss amount X is represented by

$$X = Y_l + Y_r,$$

where Y_l is the amount paid by the insurer and

$$Y_r = \begin{cases} X, & \text{if} \quad X \le d \\ d, & \text{if} \quad X > d. \end{cases}$$

In this case Y_r is called a **retained loss**, which is a loss not paid by the insurance policy.

Remark 5.4.2. For an aggregate claim S filed, a **stop-loss insurance** specifies a deductible amount d. The amount paid by the insurer can be described as

$$Y_l = \begin{cases} 0, & \text{if} \quad S \le d \\ S - d, & \text{if} \quad S > d. \end{cases}$$

A stop-loss insurance is the same form as an excess-of-loss insurance. However, the former is applied to aggregate claims, while the letter is applied to individual claims.

There are two types of deductible. If a **franchise deductible** d is applied to a loss X, then the amount paid per loss is described as

$$Y_l = \begin{cases} 0, & \text{if} \quad X \le d \\ X, & \text{if} \quad X > d. \end{cases}$$

If a **disappearing deductible** with lower d_1 and upper limit d_2 is applied to a loss X, then the amount paid per loss is described as

$$Y_l = \begin{cases} 0, & \text{if} \quad X \le d_1 \\ d_2(X - d_1)/(d_2 - d_1), & \text{if} \quad d_1 < X \le d_2 \\ X, & \text{if} \quad X > d_2. \end{cases}$$

If the loss X faces a **policy limit** u, then the insurer will pay a maximum amount of u on a claim. In this case, $Y_l = Y_c$, the amount paid per loss/claim, is denoted by

$$Y_l = \min\{X, u\} = \begin{cases} X, & \text{if} \quad X \le u \\ u, & \text{if} \quad X > u. \end{cases}$$

In this case, if $f_X(x) = 0$ for $x \le a$, then we have

$$\begin{aligned} E[Y_l] &= \int_a^u x f_X(x)dx + u(1 - F_X(u)) \\ &= a + \int_a^u (1 - F_X(x))dx. \end{aligned} \tag{5.4.1}$$

In fact, by the definition of Y_l, we see that Y_l has the pdf of X on $\{X \leq u\}$ and the pmf on $\{X > u\} = \{Y_l = u\}$. So

$$\begin{aligned} E[Y_l] &= \int_a^u x f_X(x)dx + uP(Y_l = u) \\ &= \int_a^u x f_X(x)dx + uP(X > u) \\ &= \int_a^u x f_X(x)dx + u(1 - F_X(u)). \end{aligned}$$

The right hand side equals

$$\begin{aligned} &\int_a^u x F'_X(x)dx + u - uF_X(u) \\ &= [xF_X(x)]_a^u - \int_a^u F_X(x)dx + \int_a^u 1dx + a - uF_X(u) \\ &= a + \int_a^u (1 - F_X(x))dx. \end{aligned}$$

The last equality holds because that $F_X(a) = 0$.

It is possible to combine an excess-of-loss partial insurance with a policy limit. If a deductible d and a policy limit u are applied to a loss X, then the amount paid by the insurer is described as

$$Y_l = \begin{cases} 0, & \text{if} \quad X \leq d \\ X - d, & \text{if} \quad d < X \leq u \\ u - d, & \text{if} \quad X > u. \end{cases}$$

In this case, by the similar way as above, we have

$$\begin{aligned} E[Y_l] &= \int_d^u (x - d) f_X(x)dx + (u - d)(1 - F_X(u)) \\ &= \int_d^u (1 - F_X(x))dx. \end{aligned}$$

An **insurance cap** m is a maximum claim amount that would be paid if a loss occurs on the policy. In the case that a policy limit u is applied to X, we see

$$m = \begin{cases} u, & \text{if there is no deductible} \\ u - d, & \text{if there is a deductible} \quad d. \end{cases}$$

A **proportional insurance** specifies a fraction α, where $0 < \alpha < 1$. If a loss amount X occurs, then the insurer pays the policyholder αX, i.e., $Y = Z = \alpha X$.

5.5. Reinsurance

When an insurance company, say a **ceding insurer**, can not cover all claims from the individual policies, the ceding insurer often purchases the reinsurance from a **reinsurer**. The claims covered (or paid) by the ceding insurer and reinsurer are referred to as retained the **retained claims** and **reinsured**(or **ceded**) **claims**, respectively.

A **stop-loss reinsurance** species a deductible amount d. If the aggregate claim S is less than d, the the reinsurer pays nothing and the ceding insurer pays the retained claim amount S to the policies. But if $S > d$, then the reinsurer pays the ceded claim amount $S-d$ and the ceding insurer pays the retained claim amount d. Thus we have

$$\begin{aligned} S &= (S_r : \text{retained claims}) &+& \quad (S_c : \text{ceded claims}) \\ &= \min(S, d) &+& \quad (S-d)_+ \end{aligned}$$

A **reinsurance cap** specifies a maximum amount paid by the reinsurer, say m. This would usually be set up in conjunction with a stop-loss with deductible d. In that case, the claim S_c paid by the reinsurer can be described as

$$S_c = \begin{cases} 0, & \text{if } \ S \le d \\ S-d, & \text{if } \ d < S \le d+m \\ m, & \text{if } \ S > d+m. \end{cases}$$

A **proportional reinsurance** specifies a fraction α, where $0 < \alpha < 1$. If aggregate claims of amount S occur, then the reinsurer pays αS and the ceding insurer pays $(1-\alpha)S$, i.e.,

$$S = S_r + S_c = \alpha S + (1-\alpha)S.$$

Example 5.5.1. ([**2**]) A life insurance company covers 16,000 mutually independent lives for one year term life insurance:

Class	Benefit Amount	Number covered	Probability of Claim
1	1	8000	0.025
2	2	3500	0.025
3	4	4500	0.025

The insurance company's retention limit is 2 unit per life. Reinsurance is purchases for 0.03 per unit. The ceding insurer collects a premium of $Q = E[S_r] + 2\sqrt{Var(S_r)} + R$, where S_r denotes the retained claims and R is the cost of reinsurance. Find Q.

Solution. The ceding insurer will cover all claims from classes 1 and 2, and will cover the first 2 units of claim from any policy in class 3. The ceding insurer purchases 2 units of reinsurance for each of the policies with benefit amount 4, for a total of $4,500 \times 2$=9,000 units reinsured. The cost of the reinsurance is $R = 0.03 \times 9,000 = 270$.

It remains to calculate the mean and variance of S and to calculate Q by the equation $Q = E[S_r] + 2\sqrt{Var(S_r)} + R$. For any j-th policy in Class k, since $B_j = b_k$ = constant 1 or 2 , we have $E[B_j] = b_k$ and $Var(B_j) = 0$. Thus for any j-th policy in Class k, it holds

$$E[X_j] = q_j E[B_j] = q_k b_k$$

and

$$Var(X_j) = g_j(1-q_j)E[B_j]^2 + q_j Var(B_j) = g_k(1-q_k)b_k^2.$$

For the three classes of insured individuals, we have the results given below.

k	n_k	q_k	$B_k = b_k$	$E[X_k]$	$Var(X_k)$
1	8000	0.025	1	0.025	0.024375
2	3500	0.025	2	0.050	0.097500
3	4500	0.025	2	0.050	0.097500

Then

$$\begin{aligned} E[S_r] &= \sum_{j=1}^{16,000} E[X_j] = \sum_{k=1}^{3} n_k E[X_k] \\ &= (8000)(0.025) + (3500)(0.050) + (4500)(0.050) \\ &= 600 \end{aligned}$$

and

$$\begin{aligned} Var(S_r) &= \sum_{j=1}^{16,000} Var(X_j) = \sum_{k=1}^{3} n_k Var(X_k) \\ &= (8000)(0.024375) + (3500)(0.097500) + (4500)(0.097500) \\ &= 975 = 16^2. \end{aligned}$$

Thus, the total premium that the insurer collects is

$$\begin{aligned} Q &= E[S_r] + 2\sqrt{Var(S_r)} + R \\ &= 600 + 2\sqrt{975} + 270 \\ &= 932.45. \end{aligned}$$

Part 2

SOA/CAS Sample Questions

Introduction

In this part we discuss 106 questions which are cited from home site on the World Wide Web,

http://www.soa.org

(see [4]). This part emphasizes to classify these questions by subjects and give more heuristic solutions, than original ones, in which the the followings can be found

- mathematical meaning for given question,
- assumptions that we are given,
- key tool to solve the question,
- detail calculation.

The author thanks the publishers of Society of Actuaries for permission to include 106 questions from SOA/CAS Exam P Sample Questions in this textbook.

CHAPTER 1

Introduction to Probability

1.1. Sample spaces and events

Question 1.1.1. ([4], No. 5) An auto insurance company has 10,000 policyholders. Each policyholder is classified as

(i) young or old;

(ii) male or female; and

(iii) married or single.

Of these policyholders, 3,000 are young, 4,600 are male, and 7,000 are married. The policyholders can also be classified as 1,320 young males, 3,010 married males, and 1,400 young married persons. Finally, 600 of the policyholders are young married males.

How many of the company's policyholders are young, female, and single?

SOLUTION We denote

$$
\begin{aligned}
Y &= \{\text{young policyholders}\},\\
O &= \{\text{old policyholders}\},\\
M &= \{\text{males}\},\\
F &= \{\text{females}\},\\
S &= \{\text{single policyholders}\},\\
S^c &= \{\text{married policyholders}\}.
\end{aligned}
$$

Let $N(C)$ be the number of the policyholders in classification C. The problem is to find $N(Y \cap F \cap S)$. To do this, we use the equality

$$N(A \cap B) = N(A) - N(A \cap B^c)$$

for any classifications A and B.

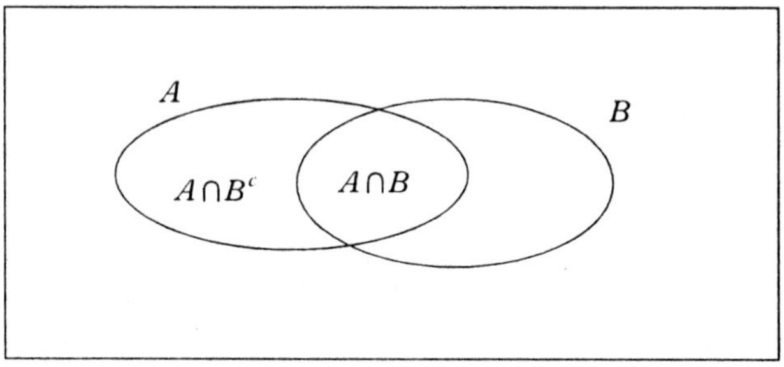

$$\begin{aligned} N(Y \cap F \cap S) &= N(Y \cap F) - N(Y \cap F \cap S^c) \\ &= N(Y) - N(Y \cap M) - \{N(Y \cap S^c) - N(Y \cap S^c \cap M)\} \\ &= 3,000 - 1,320 - (1,400 - 600) \\ &= 880. \end{aligned}$$

1.2. The axioms of probability

Question 1.2.1. ([4], No. 1) A survey of a group's viewing habits over the last year revealed the following information:

(i) 28% watched gymnastics

(ii) 29% watched baseball

(iii) 19% watched soccer

(iv) 14% watched gymnastics and baseball

(v) 12% watched baseball and soccer

(vi) 10% watched gymnastics and soccer

(vii) 8% watched all three sports.

Calculate the percentage of the group that watched none of the three sports during the last year.

SOLUTION. Let G be an event that a viewer watched gymnastics over the last year, B an event that a viewer watched baseball, and S an event that a viewer watched soccer. The question is to find $100P(G^c \cap B^c \cap S^c)$. We are given

$$\begin{aligned}
&P(G) = 0.28, \;\; P(B) = 0.29, \;\; P(S) = 0.19,\\
&P(G \cap B) = 0.14,\\
&P(B \cap S) = 0.12,\\
&P(S \cap G) = 0.10,\\
&P(G \cap B \cap S) = 0.08
\end{aligned}$$

To solve this question, we use Remark 1.2.5.

$$\begin{aligned}
P(G^c \cap B^c \cap S^c) &= P((G \cup B \cup S)^c)\\
&= 1 - P(G \cup B \cup S)\\
&= 1 - \{P(G) + P(B) + P(S)\\
&\qquad - P(G \cap B) - P(B \cap S) - P(S \cap G)\\
&\qquad + P(G \cap B \cap S)\}\\
&= 1 - (0.28 + 0.29 + 0.19 - 0.14 - 0.12 - 0.10 + 0.08)\\
&= 1 - 0.48\\
&= 0.52.
\end{aligned}$$

So the answer is 52.

Question 1.2.2. ([4], No. 3) You are given $P(A \cup B) = 0.7$ and $P(A \cup B^c) = 0.9$. Determine $P(A)$.

SOLUTION. To solve this problem we use Theorem 1.2.4. Since

$$\begin{aligned} P(A) &= P(A \cap B) + P(A \cap B^c) \\ &= P(A) + P(B) - P(A \cup B) + P(A) + P(B^c) - P(A \cup B^c) \\ &= 2P(A) + 1 - 0.7 - 0.9 \\ &= 2P(A) - 0.6, \end{aligned}$$

it holds that $P(A) = 0.6$.

Question 1.2.3. ([4], No. 8) Among a large group of patients recovering from shoulder injuries, it is found that 22% visit both a physical therapist and a chiropractor, whereas 12% visit neither of these. The probability that a patient visits a chiropractor exceeds by 0.14 the probability that a patient visits a physical therapist.

Determine the probability that a randomly chosen member of this group visits a physical therapist.

SOLUTION. Let C be an event that a patient visits a chiropractor and T an event that a patient visits a therapist. The question is to find $P(T)$. We are given

$$\begin{aligned} P(C \cap T) &= 0.22, \\ P(C^c \cap T^c) &= 0.12, \\ P(C) &= P(T) + 0.14. \end{aligned}$$

By Theorem 1.2.4, we have

$$\begin{aligned} P(C \cup T) &= P(C) + P(T) - P(C \cap T) \\ &= P(T) + 0.14 + P(T) - 0.22 \\ &= 2P(T) + 0.36. \end{aligned}$$

On the other hand,

$$\begin{aligned}P(C \cup T) &= 1 - P((C \cup T)^c)\\ &= 1 - P(C^c \cap T^c)\\ &= 1 - 0.12\\ &= 0.88.\end{aligned}$$

So we have

$$2P(T) + 0.36 = 0.88$$

and $P(T) = 0.48$.

1.3. Combinatorial principles - permutations and combinations

1.4. Conditional probability and independence

Question 1.4.1. ([4], No. 4) An urn contains 10 balls: 4 red and 6 blue. A second urn contains 16 red balls and an unknown number on blue balls. A single ball is drown from each urn. The probability that both balls are the same color is 0.44.

Calculate the number of blue balls in the second urn.

SOLUTION. Let R_i be an event that a red ball drown from urn i and B_i an event that a red ball drown from urn i, $i = 1, 2$. The problem is to find the number x of blue balls in urn 2. From the conditions, we have

$$\begin{aligned}0.44 &= P\left((R_1 \cap R_2) \cup (B_1 \cap B_2)\right)\\ &= P(R_1 \cap R_2) + P(B_1 \cap B_2)\\ &= P(R_1)P(R_2) + P(B_1)P(B_2)\\ &= \left(\frac{4}{10}\right)\left(\frac{16}{16+x}\right) + \left(\frac{6}{10}\right)\left(\frac{x}{16+x}\right).\end{aligned}$$

So we have

$$44(16+x) = 40 \times 16 + 60x$$

and $x = 4$.

Question 1.4.2. ([4], No. 6) A public health researcher examines the medical records of a group of 937 men who died in 1999 and discovers that 210 of the men died from causes related to heart disease. Moreover, 312 of the 937 men had at least one parent who suffered from heart disease, and, of these 312 men, 102 died from causes related to heart disease.

Determine the probability that a man randomly selected from this group died of cause related to heart disease, given that neither of his parents suffered from heart disease.

SOLUTION. Let H be an event that a death is due to heart disease, F an event that at least one parent suffered from heart disease, and S the set of 937 men who died in 1999. The problem is to determine $P(H|F^c)$. By the definition of the conditional probability, it holds

$$P(H|F^c) = \frac{P(H \cap F^c)}{P(F^c)}.$$

Here

$$\begin{aligned} P(F^c) &= 1 - P(F) = 1 - \frac{N(F)}{N(S)} \\ &= 1 - \frac{312}{937} = \frac{625}{937} \end{aligned}$$

and

$$\begin{aligned} P(H \cap F^c) = P(H) - P(H \cap F) &= \frac{N(H)}{N(S)} - \frac{N(H \cap F)}{N(S)} \\ &= \frac{210}{937} - \frac{102}{937} = \frac{108}{937}. \end{aligned}$$

Hence

$$P(H|F^c) = \frac{108/937}{625/937} = \frac{108}{625} = 0.173.$$

Question 1.4.3. ([4], No. 7) An insurance company estimates that 40% of policyholders who have only an auto policy will renew next year and 60% of policyholders who have only a homeowners policy will renew next. The company estimates that 80% of policyholders who have both an auto and homeowners will renew at least one of those policies next year.
Company records show that 65% of policyholders have an auto policy, 50% of policyholders have a homeowners policy, and 15% of policyholders who have both an auto and homeowners policy.

Using the company's estimates, calculate the percentage of policyholders that will renew at least one policy next year.

SOLUTION. Let A be an event that a policyholder has an auto policy, H an event that a policyholder has a homeowners policy, and R an event that a policyholder will renew at least one policy next year. The problem is to calculate $P(R)$. From the company's estimates, we have

$$\begin{aligned}&P(R|A \cap H^c) = 0.4,\ P(R|A^c \cap H) = 0.6,\ P(R|A \cap H) = 0.8,\\&P(A) = 0.65,\ P(H) = 0.5,\\&P(A \cap H) = 0.15.\end{aligned}$$

Since

$$S = A \cup H = (A^c \cap H) \cup (A \cap H) \cup (A \cap H^c),$$

we can see that

$$P(A^c \cap H) = 0.35$$

and

$$P(A \cap H^c) = 0.5.$$

By the theorem of total probability, i.e. (1.4.1),

$$\begin{aligned} P(R) &= P(R|A\cap H^c)P(A\cap H^c) + P(R|A^c\cap H)P(A^c\cap H) \\ &\qquad + P(R|A\cap H)P(A\cap H) \\ &= (0.4)(0.5) + (0.6)(0.35) + (0.8)(0.15) \\ &= 0.53 \end{aligned}$$

Thus the answer is 53.

Question 1.4.4. ([4], No. 9) An insurance company examines its pool of auto insurance customers and gathers the following information:

(i) All customers insure at least one car.

(ii) 70% of the customers insure more than one car.

(iii) 20% of the customers insure a sports car.

(iv) Of those customers who insure more than one car, 15% insure a sports car.

Calculate the probability that a randomly selected customer insures exactly one car and that car is not a sports car.

SOLUTION. Let A be an event that a customer insures more than one car and B an event that a customer insures a sports car. The question is to calculate $P(A^c\cap B^c)$. We are given

$$P(A) = 0.7,\ \ P(B) = 0.2,$$
$$P(B|A) = 0.15.$$

By Theorem 1.2.4 and the definition of the conditional probability, we see

$$\begin{aligned} P(A^c\cap B^c) &= 1 - P(A\cup B) \\ &= 1 - \{P(A) + P(B) - P(A\cap B)\} \\ &= 1 - \{P(A) + P(B) - P(B|A))P(A)\} \\ &= 1 - (0.7 + 0.2 - (0.15)(0.7) \\ &= 0.205. \end{aligned}$$

Question 1.4.5. ([4], No. 11) An actuary studying the insurance preferences of automobile owners makes the following conclusions:

(i) An automobile owner is twice as likely to purchase collision coverage as disability coverage.

(ii) The event that an automobile owner purchases collision coverage is independent of the event that he or she purchases disability coverage.

(iii) The probability that an automobile owner purchases both collision and disability coverage is 0.15.

What is the probability that an automobile owner purchases neither collision nor disability coverage?

SOLUTION. Let C be an event that a policyholder purchases collision coverage and D an event that a policyholder purchases disability coverage. The question is to calculate $P(C^c \cap D^c)$. We are given

$$\begin{aligned} &C, D : \text{independent}, \\ &P(C) = 2P(D), \\ &P(C \cap D) = 0.15. \end{aligned}$$

To solve this, we also use Theorem 1.2.4.

$$\begin{aligned} P(C^c \cap D^c) &= P((C \cup D)^c) \\ &= 1 - P(C \cup D) \\ &= 1 - \{P(C) + P(D) - P(C \cap D)\} \\ &= 1 - \{3P(D) - 0.15\} \\ &= 1.15 - 3P(D). \end{aligned}$$

Since $0.15 = P(C \cap D) = P(C)P(D) = 2P(D)^2$, we have

$$P(D) = \sqrt{\frac{0.15}{2}} = \sqrt{0.075}.$$

Hence

$$P(C^c \cap D^c) = 1.15 - 3\sqrt{0.075} = 1.15 - 3\sqrt{0.275^2} = 0.325.$$

Question 1.4.6. ([4], No. 17) An insurance company pays hospital claims. The number of claims that include emergence room or operating room charge is 85% of the total number of claims. The number of claims that do not include emergency room charges is 25% of the total number of claims. The occurrence of emergency room charges is independent of the occurrence of operating room charges on hospital claims.

Calculate the probability that a claim submitted to the insurance company includes operating room charges.

SOLUTION. Let O and E be an event that a claim includes operating room charges and emergency room charges, respectively. The question is to calculate $P(O)$. We are given

$$\begin{aligned} &P(O \cup E) = 0.85, \\ &P(E^c) = 0.25, \\ &O, E : \text{independent}. \end{aligned}$$

By the condition,

$$\begin{aligned} P(O \cup E) &= P(O) + P(E) - P(O \cap E) \\ &= P(O) + P(E) - P(O)P(E) \\ &= P(O) + 0.75 - 0.75P(O). \\ &= 0.25P(O) + 0.75. \end{aligned}$$

But since $P(O \cup E) = 0.85$, we have

$$0.25P(O) + 0.75 = 0.85$$

and so $P(O) = 0.40$.

Question 1.4.7. ([4], No. 20) An insurance company issues life insurance policies in three separate categories: standard, preferred, and ultra-preferred. Of the company's policyholders, 50% are standard, 40% are preferred, and 10% are ultra-preferred. Each standard policyholder has probability 0.010 of dying in the next year, each preferred policyholder has probability 0.005 of dying in the next year, and each ultra-preferred policyholder has probability 0.001 of dying in the next year.

A policyholder dies in the next year. What is the probability that the deceased policyholder was ultra-preferred?

SOLUTION. Let

$$\begin{aligned}
S &= \text{event that a policyholder is standard,}\\
F &= \text{event that a policyholder is preferred,}\\
U &= \text{event that a policyholder is ultra-preferred,}\\
D &= \text{event that a policyholder dies in the next year.}
\end{aligned}$$

The question is to find $P(U|D)$. We are given

$$\begin{aligned}
&P(S) = 0.5,\ P(F) = 0.4,\ P(U) = 0.1,\\
&P(D|S) = 0.010,\\
&P(D|F) = 0.005,\\
&P(D|U) = 0.001.
\end{aligned}$$

By Bayes' rule,

$$\begin{aligned}
P(U|D) &= \frac{P(D|U)P(U)}{P(D|S)P(S) + P(D|F)P(F) + P(D|U)P(U)}\\
&= \frac{(0.001)(0.1)}{(0.010)(0.5) + (0.005)(0.4) + (0.001)(0.1)}\\
&= 0.0141.
\end{aligned}$$

Question 1.4.8. ([4], No. 21) Upon arrival at a hospital's emergency room, patients are categorized according to their condition as critical, serious, or stable. In the past year:

(i) 10% of the emergency room patients were critical;

(ii) 30% of the emergency room patients were serious;

(iii) the rest of the emergency room patients were stable;

(iv) 40% of the critical patients died;

(v) 10% of the serious patients died; and

(vi) 1% of the stable patients died.

Given that a patients survived, what is the probability that the patients was categorized as serious upon arrival?

SOLUTION. Let

$$\begin{aligned} A &= \text{event that an emergency room patient is critical,} \\ B &= \text{event that an emergency room patient is serious,} \\ C &= \text{event that an emergency room patient is stable,} \\ D &= \text{event that an emergency room patient die.} \end{aligned}$$

The question is to find $P(B|D^c)$. We are given

$$\begin{aligned} &P(A) = 0.1, \ P(B) = 0.3, \ P(C) = 0.6, \\ &P(D^c|A) = 1 - 0.4 = 0.6, \\ &P(D^c|B) = 1 - 0.1 = 0.9, \\ &P(D^c|C) = 1 - 0.01 = 0.99. \end{aligned}$$

By Bayes' rule,

$$\begin{aligned} P(B|D^c) &= \frac{P(D^c|B)P(B)}{P(D^c|A)P(A) + P(D^c|B)P(B) + P(D^c|C)P(C)} \\ &= \frac{(0.9)(0.3)}{(0.6)(0.1) + (0.9)(0.3) + (0.99)(0.6)} \\ &= 0.29. \end{aligned}$$

Question 1.4.9. ([4], No. 22) A health study tracked a group of persons for five years. At the beginning of the study, 20% were classified as heavy smokers, 30% as light smokers, and 50% as nonsmokers.

Results of the study showed that light smokers were twice as likely as nonsmokers to die during the five-year study, but only half as likely as heavy smokers.

A randomly selected participant from the study died over the five-year period. Calculate the probability that the participant was a heavy smoker.

SOLUTION. Let

H = event that a participant is a heavy smoker,

L = event that a participant is a light smoker,

N = event that a participant is a nonsmoker,

D = event that a participant die over the five-year period.

The question is to find $P(H|D)$. Results of the study shows

$$P(H) = 0.2, \ P(L) = 0.3, \ P(N) = 0.5,$$

$$P(D|L) = 2P(D|N) = \frac{1}{2}P(D|H).$$

By Bayes' rule,

$$\begin{aligned} P(H|D) &= \frac{P(D|H)P(H)}{P(D|H)P(H) + P(D|L)P(L) + P(D|N)P(N)} \\ &= \frac{P(D|H)P(H)}{P(D|H)P(H) + 1/2P(D|H)P(L) + 1/4P(D|H)P(N)} \\ &= \frac{P(H)}{P(H) + 1/2P(L) + 1/4P(N)} \\ &= \frac{0.2}{0.2 + (1/2)(0.3) + (1/4)(0.5)} \\ &= 0.42. \end{aligned}$$

Question 1.4.10. ([1], No. 25) A blood test indicates the presence of a particular disease 95% of the time when the disease is actually present. The same test indicates the presence of the disease 0.5% of the time when the disease is not present. One percent of the population actually has the disease.

Calculate the probability that a person has the disease given that the test indicates the presence of the disease.

SOLUTION. Let

$$Y = \text{event that the test indicates the presence of the disease,}$$
$$D = \text{event that the disease is actually present.}$$

The question is to calculate $P(D|Y)$. We are given

$$P(Y|D) = 0.95,$$
$$P(Y|D^c) = 0.005,$$
$$P(D) = 0.01$$

By Bayes' rule,

$$\begin{aligned} P(D|Y) &= \frac{P(Y|D)P(D)}{P(Y|D)P(D) + P(Y|D^c)P(D^c)} \\ &= \frac{(0.95)(0.01)}{(0.95)(0.01) + (0.005)(0.99)} \\ &= 0.657. \end{aligned}$$

Question 1.4.11. ([4], No. 26) The probability that a randomly chosen male has a circulation problem is 0.25. Males who have a circulation problem are twice as likely to be smokers as those who do not have a circulation problem.

What is the conditional probability that a male has a circulation problem, given that he is a smoker?

SOLUTION. Let M be an event that a male is a smoker and C an event that a male has a circulation problem. We are given

$$P(C) = 0.25,$$
$$P(M|C) = 2P(M|C^c).$$

The question is to find $P(C|M)$. By Bayes' rule,

$$\begin{aligned}
P(C|M) &= \frac{P(M|C)P(C)}{P(M|C)P(C) + P(M|C^c)P(C^c)} \\
&= \frac{P(M|C)P(C)}{P(M|C)P(C) + 1/2P(M|C)P(C^c)} \\
&= \frac{P(C)}{P(C) + 1/2P(C^c)} \\
&= \frac{0.25}{0.25 + (1/2)(0.75)} \\
&= \frac{2}{5}.
\end{aligned}$$

Question 1.4.12. ([4], No. 27) A study of automobile accidents produced the following data:

Model year	Proportion of all vehicle	Probability of involvement in an accident
1997	0.16	0.05
1998	0.18	0.02
1999	0.20	0.03
Other	0.46	0.04

An automobile from one of the model years 1997, 1998, 1999 was involved in an accident.

Determine the probability that the model year of this automobile is 1997.

SOLUTION. Let

$U =$ event that the model year is 1997,

$V =$ event that the model year is 1998,

$W =$ event that the model year is 1999,

$A =$ event that an automobile from one of the model years 1997, 1998, and 1999 involves in an accident.

The study of automobile accident shows

$$P(U) = 0.16, \ P(V) = 0.18, \ P(W) = 0.20,$$

$$P(A|U) = 0.05, \ P(A|V) = 0.02, \ P(A|W) = 0.03$$

The question is to determine $P(U|A)$. By Bayes' rule,

$$\begin{aligned} P(U|A) &= \frac{P(A|U)P(U)}{P(A|U)P(U) + P(A|V)P(V) + P(A|W)P(W)} \\ &= \frac{(0.05)(0.16)}{(0.05)(0.16) + (0.02)(0.18) + (0.03)(0.20)} \\ &= 0.45. \end{aligned}$$

CHAPTER 2

Random Variables and Their Probability Distributions

2.1. Random variables and distributions

Question 2.1.1. ([4], No. 14) In modeling the number of claims filed by an individual under an automobile policy during a three-year period, an actuary makes the simplifying assumption that for all integers $n \geq 0$, $p_{n+1} = \frac{1}{5}p_n$, where p_n represents the probability that the policyholder files n claims during the period.

Under this assumption, what is the probability that a policyholder files more than one claim during the period?

Solution. Let N be the number of claims. Then $p_k = P(N = k), k = 0, 1, 2, \cdots$. The question is to find $P(N > 1)$. Since, for any $k = 1, 2, \cdots$,

$$p_k = \frac{1}{5}p_{k-1} = \left(\frac{1}{5}\right)\left(\frac{1}{5}\right)p_{k-2} = \cdots = \left(\frac{1}{5}\right)^k p_0,$$

we have

$$1 = \sum_{k=0}^{\infty} p_k = \sum_{k=0}^{\infty} \left(\frac{1}{5}\right)^k p_0 = \frac{1}{1 - 1/5}p_0 = \frac{5}{4}p_0$$

and hence

$$p_0 = \frac{4}{5}.$$

Thus

$$\begin{aligned}
P(N > 1) &= 1 - P(N \leq 1) \\
&= 1 - P(N = 0 \text{ or } N = 1) \\
&= 1 - \{P(N = 0) + P(N = 1)\} \\
&= 1 - (p_0 + p_1) \\
&= 1 - \left\{\frac{4}{5} + \left(\frac{1}{5}\right)\left(\frac{4}{5}\right)\right\} \\
&= \frac{1}{25}
\end{aligned}$$

Question 2.1.2. ([4], No. 16) An insurance company determines that N, the number of claims received in a week, is a random variable with $P(N = n) = 1/2^{n+1}$, where $n \geq 0$. The company also determines that the number of claims received in a given week is independent of the number of claims received in any other week.

Determine the probability that exactly seven claims will be received during a given two-week period.

SOLUTION. Let

N_1 : number of claims received in the first week,

N_2 : number of claims received in the second week.

By assumption, for any $j = 1, 2$,

$$P(N_j = n) = \frac{1}{2^{n+1}}.$$

The question is to determine $P(N_1 + N_2 = 7)$. By the theorem of total probability (1.4.1), we have

$$\begin{aligned}
P(N_1 + N_2 = 7) &= \sum_{n=0}^{7} P(N_1 + N_2 = 7 | N_1 = n) P(N_1 = n) \\
&= \sum_{n=0}^{7} P(N_1 + N_2 = 7, N_1 = n) \\
&= \sum_{n=0}^{7} P(N_1 = n, N_2 = 7 - n) \\
&= \sum_{n=0}^{7} \left(\frac{1}{2^{n+1}}\right)\left(\frac{1}{2^{8-n}}\right) \\
&= \sum_{n=0}^{7} \frac{1}{2^9} \\
&= 8 \times \frac{1}{2^9} = \frac{1}{2^6} = \frac{1}{64}.
\end{aligned}$$

Question 2.1.3. ([**4**], No. 24) The number of injury claims per month is modeled by a random variable N with

$$P(N = n) = \frac{1}{(n+1)(n+2)}, \quad n \geq 0.$$

Determine the probability of at least one claim during a particular month, given that there have been at most four claims during that month.

SOLUTION. The question is to determine the conditional probability $P(N \geq 1|N \leq 4)$. By its definition, we get

$$\begin{aligned}
&P(N \geq 1|N \leq 4) \\
&= \frac{P(1 \leq N \leq 4)}{P(N \leq 4)} \\
&= \frac{P(N=1 \text{ or } N=2 \text{ or } N=3 \text{ or } N=4)}{P(N=0 \text{ or } N=1 \text{ or } N=2 \text{ or } N=3 \text{ or } N=4)} \\
&= \frac{P(N=1)+P(N=2)+P(N=3)+P(N=4)}{P(N=0)+P(N=1)+P(N=2)+P(N=3)+P(N=4)} \\
&= \frac{\frac{1}{6}+\frac{1}{12}+\frac{1}{20}+\frac{1}{30}}{\frac{1}{2}+\frac{1}{6}+\frac{1}{12}+\frac{1}{20}+\frac{1}{30}} \\
&= \frac{10+5+3+2}{30+10+5+3+2} = \frac{2}{5}.
\end{aligned}$$

Question 2.1.4. ([4], No. 34) The lifetime X of a machine part has a continuous distribution on the interval (0,40) with probability density function $f_X(x)$, where $f_X(x)$ is proportional to $(10+x)^{-2}$.

Calculate the probability that the life time of the machine part is less than 6.

SOLUTION. The question is to calculate $P(X < 6)$. Clearly the probability density function has the form

$$f_X(x) = \begin{cases} C(10+x)^{-2} & \text{if } \ 0 < x < 40, \\ 0 & \text{otherwise} \end{cases}$$

for some constant C. Let us, first, determine C by the property of the probability density function.

$$\begin{aligned}
1 &= \int_{-\infty}^{\infty} f_X(x)dx = \int_0^{40} C(10+x)^{-2}dx \\
&= \left[-\frac{C}{10+x}\right]_{x=0}^{x=40} = -\frac{C}{50} + \frac{C}{10} = \frac{2C}{25}.
\end{aligned}$$

So we get $C = 25/2$. Then we can see that

$$\begin{aligned} P(X < 6) &= \int_0^6 \frac{25}{2}(10+x)^{-2}dx \\ &= \left[-\frac{C}{10+x}\right]_0^6 \\ &= -\frac{25}{32} + \frac{25}{20} = 0.47. \end{aligned}$$

Question 2.1.5. ([4], No. 36) A group insurance policy covers the medical claims of employees of a small company. The value, V, of the claims made in one year is described by

$$V = 100,000Y$$

where Y is a random variable with density function

$$f_V(y) = \begin{cases} k(1-y)^4 & \text{for } \ 0 < y < 1, \\ 0 & \text{otherwise,} \end{cases}$$

where k is a constant.

What is the conditional probability that V exceeds 40,000, given that V exceeds 10,000?

Solution. The question is to calculate $P(V > 40,000|V > 10,000)$. Let us determine the constant k. From

$$1 = \int_0^1 k(1-y)^4 dy = \left[-k\frac{1}{5}(1-y)^5\right]_0^1 = \frac{k}{5},$$

we get $k = 5$. By the definition of the conditional probability,

$$\begin{aligned} P(V > 40,000|V > 10,000) &= \frac{P(V > 40,000 \text{ and } V > 10,000)}{P(V > 10,000)} \\ &= \frac{P(V > 40,000)}{P(V > 10,000)}. \end{aligned}$$

Here

$$\begin{aligned} P(V > 40,000) &= P(100,000Y > 40,000) \\ &= P(Y > 0.4) \\ &= \int_{0.4}^{1} 5(1-y)^4 dy = \left[-(1-y)^5\right]_{0.4}^{1} \\ &= (1-0.4)^5 = 0.0078. \end{aligned}$$

By the same calculation, we have

$$P(V > 10,000) = P(Y > 0.1) = (1-0.1)^5 = 0.59.$$

Thus

$$P(V > 40,000 | V > 10,000) = \frac{0.0078}{0.59} = 0.132.$$

Question 2.1.6. ([4], No. 38) An insurance company insures a large number of homes. The insured value, X, of a randomly selected home is assumed to follow a distribution with density function

$$f_X(x) = \begin{cases} 3x^{-4} & \text{for} \quad x > 1, \\ 0 & \text{otherwise.} \end{cases}$$

Given that a randomly selected home is insured for at least 1.5, what is the probability that it is insured for less than 2?

SOLUTION. The question is to calculate the conditional probability $P(X < 2 | X \geq 1.5)$. By the definition of the conditional probability, we have

$$\begin{aligned}P(X < 2|X \geq 1.5) &= \frac{P(1.5 \leq X < 2)}{P(X \geq 1.5)} \\ &= \frac{P(1.5 \leq X < 2)}{1 - P(X < 1.5)} \\ &= \frac{\int_{1.5}^{2} 3x^{-4}dx}{1 - \int_{1}^{1.5} 3x^{-4}dx} \\ &= \frac{[-1/x^3]_{1.5}^{2}}{1 - [-1/x^3]_{1}^{1.5}} \\ &= 0.578\end{aligned}$$

Question 2.1.7. ([4], No. 71) The time, T, that a manufacturing system is out of operation has cumulative distribution function

$$F_T(t) = \begin{cases} 1 - \left(\frac{2}{t}\right)^2 & \text{for} \quad t > 2, \\ 0 & \text{otherwise.} \end{cases}$$

The resulting cost to the company is $Y = T^2$.

Determine the density function of Y, for $y > 4$.

SOLUTION. Let $F_Y(y)$ and $f_Y(y)$ be the cumulative distribution function and density function of Y, respectively. To solve this question, we use the relation between $F_Y(y)$ and $f_Y(y)$ as following

$$f_Y(y) = \frac{dF_Y(y)}{dy}.$$

Let us, first, find $F_Y(y)$.

$$\begin{aligned}F_Y(y) &= P(Y \leq y) = P(T^2 \leq y) = P(T \leq \sqrt{y}) \\ &= F_T(\sqrt{y}) = 1 - \left(\frac{2}{\sqrt{y}}\right)^2 \\ &= 1 - \frac{4}{y}.\end{aligned}$$

Hence

$$f_Y(y) = F'_Y(y) = \left(1 - \frac{4}{y}\right)' = \frac{4}{y^2}.$$

Question 2.1.8. ([4], No. 75) The monthly profit X_1 of Company I can be modeled by a continuous random variable with density function f_1. Company II has a monthly profit that is twice that of Company I.

Determine the probability density function of the monthly profit of Company II.

SOLUTION. Let f_2 be the probability density function of the monthly profit X_2 of Company II. The question is to give a representation of f_2 by f_1. First let us find a relation between the distribution functions F_1 and F_2 of X_1 and X_2, respectively.

$$\begin{aligned} F_2(x) &= P(X_2 \leq x) = P(2X_1 \leq x) \\ &= P(X_1 \leq x/2) \\ &= F_1(x/2). \end{aligned}$$

Thus

$$f_2(x) = F_2'(x) = [F_1(x/2)]' = \frac{1}{2}F_1'(x/2) = \frac{1}{2}f_1(x/2).$$

2.2. Expectation and variance

Question 2.2.1. ([4], No. 44) An insurance policy pays 100 per day for up to 3 days of hospitalization and 50 per day for each day of hospitalization thereafter.

The number of days of hospitalization, X, is a discrete random variable with probability mass function

$$P(X = k) = \begin{cases} \frac{6-k}{15} & \text{for} \quad k = 1, 2, 3, 4, 5, \\ 0 & \text{otherwise.} \end{cases}$$

Determine the expected payment for hospitalization under this policy.

SOLUTION. Let $Y = g(X)$ be the insurance payment for the number X of days of hospitalization. Then

$$Y = g(X) = \begin{cases} 100X & \text{for} \quad k = 1, 2, 3, \\ 300 + 50(X - 3) & \text{for} \quad k = 4, 5. \end{cases}$$

The question is to determine $E[Y]$. By Theorem 2.2.4,

$$\begin{aligned} E[Y] &= \sum_{k=1}^{5} g(k)P(X = k) \\ &= (100)(1)\left(\frac{6-1}{15}\right) + (100)(2)\left(\frac{6-2}{15}\right) + (100)(3)\left(\frac{6-3}{15}\right) \\ &\quad + (300 + 50(4-3))\left(\frac{6-4}{15}\right) + (300 + 50(5-3))\left(\frac{6-5}{15}\right) \\ &= \frac{1}{15}(600 + 800 + 900 + 700 + 400) \\ &= \frac{3300}{15} = 220. \end{aligned}$$

Question 2.2.2. ([4], No. 45) Let X be a continuous random variable with density function

$$f_X(x) = \begin{cases} \frac{|x|}{10} & \text{for} \quad -2 \le x \le 4 \\ 0 & \text{otherwise.} \end{cases}$$

Calculate the expected value of X.

SOLUTION. The question is to calculate $E[X]$. The density function is rewritten as follows;

$$f_X(x) = \begin{cases} \frac{-x}{10} & \text{for} \quad -2 \le x < 0 \\ \frac{x}{10} & \text{for} \quad 0 \le x \le 4 \\ 0 & \text{otherwise.} \end{cases}$$

Then we have

$$\begin{aligned}
E[X] &= \int_{-\infty}^{\infty} x f_X(x)dx \\
&= \int_{-2}^{0} x\frac{-x}{10}dx + \int_{0}^{4} x\frac{x}{10}dx \\
&= \int_{-2}^{0} -\frac{x^2}{10}dx + \int_{0}^{4} \frac{x^2}{10}dx \\
&= \left[-\frac{x^3}{30}\right]_{-2}^{0} + \left[\frac{x^3}{30}\right]_{0}^{4} \\
&= -\frac{8}{30} + \frac{64}{30} \\
&= \frac{28}{15}.
\end{aligned}$$

Question 2.2.3. ([4], No. 48) An insurance policy on an electrical device pays a benefit of 4,000 if the device fails during the first year. The amount of the benefit decreases by 1,000 each successive year until it reaches 0. If the device has not failed by the beginning of any year, the probability of failure during that year is 0.4.

What is the expected benefit under this policy?

Solution. Let

$$X : \text{the year that the device fails}$$
$$Y : \text{the benefit amount for } X.$$

The question is to find $E[Y]$. We are given that

$$p_X(k) = P(X = k) = (0.6)^{k-1}(0.4) \quad \text{for} \quad k = 1, 2, \cdots$$

and

$$Y = g(X) = \begin{cases} 1,000(5 - X) & \text{for} \quad X = 1, 2, 3, 4 \\ 0 & \text{otherwise.} \end{cases}$$

Then we have

$$
\begin{aligned}
E[Y] = E[g(X)] &= \sum_{k=1}^{4} g(k)p_X(k) \\
&= 4,000(0.4) + 3,000(0.6)(0.4) \\
&\qquad + 2,000(0.6)^2(0.4) + 1,000(0.6)^3(0.4) \\
&= 2694.
\end{aligned}
$$

Question 2.2.4. ([4], No. 55) An insurance company'monthly claims are modeled by a continuous positive random variable X, whose probability density function is proportional to $(1+x)^{-4}$, where $0 < x < \infty$.

Determine the company's expected monthly claims.

SOLUTION. The probability density function of X is the form

$$f_X(x) = \frac{K}{(1+x)^4}, \quad 0 < x < \infty$$

for some constant K. To determine K, note that

$$1 = \int_{-\infty}^{\infty} f_X(x)dx = \int_0^{\infty} \frac{K}{(1+x)^4} dx = \frac{K}{3}.$$

Then we have $K = 3$ and so

$$f_X(x) = \frac{3}{(1+x)^4}, \quad 0 < x < \infty.$$

By the definition of the expectation, we see

$$E[X] = \int_{-\infty}^{\infty} x f_X(x)dx = \int_0^{\infty} \frac{3x}{(1+x)^4} dx.$$

To find the the definite integral in the right hand side, we let $u = 1 + x$. Note that $dx = du$, $u(0) = 1$ and $u(\infty) = \infty$. Then we have

$$\begin{aligned} E[X] &= \int_1^\infty \frac{3(u-1)}{u^4} du \\ &= 3 \int_1^\infty \left(u^{-3} - u^{-4}\right) du \\ &= 3 \left[\frac{u^{-2}}{-2} - \frac{u^{-3}}{-3}\right]_1^\infty \\ &= \frac{3}{2} - 1 = \frac{1}{2}. \end{aligned}$$

Question 2.2.5. ([4], No. 57) An actuary determines that the claim size for a certain class of accidents is a random variable, X, with moment generating function

$$M_X(t) = \frac{1}{(1 - 2,500t)^4}.$$

Determine the standard deviation of the claim size for this class of accidents.

SOLUTION. The question is to determine $\sqrt{Var(X)}$. By Remark 2.2.7(4), we see

$$\begin{aligned} E[X] = M_X'(0) &= \left.\frac{(-4)(-2,500)}{(1-2,500t)^5}\right|_{t=0} = 10,000, \\ E[X^2] = M_X''(0) &= \left.\frac{(-5)(-4)(-2,500)^2}{(1-2,500t)^6}\right|_{t=0} = 125,000,000. \end{aligned}$$

From these we have

$$\begin{aligned} Var(X) &= E[X^2] - E[X]^2 \\ &= 125,000,000 - (10,000)^2 = 25,000,000 \end{aligned}$$

and so $\sqrt{Var(X)} = 5,000$.

Question 2.2.6. ([4], No. 58) A company insures homes in three cities, J, K, and L. Since sufficient distance separates the cities, it is reasonable to assume that the losses occurring in these cities are independent.

The moment generating functions for the loss distributions of the cities are:

$$M_J(t) = (1-2t)^{-3}$$
$$M_K(t) = (1-2t)^{-2.5}$$
$$M_L(t) = (1-2t)^{-4.5}.$$

Let X represent the combined losses from the three cities.

Calculate $E[X^3]$.

SOLUTION. Let X_J, X_K and X_L be the losses occurring is the cities, J, K, and L, respectively. Then

$$X = X_J + X_K + X_L.$$

The question is to calculate $E[X^3]$. To do this, let us find the moment generating function $M_X(t)$ of X.

$$\begin{aligned}
M_X(t) &= E[e^{tX}] = E\left[e^{t(X_J+X_K+X_L)}\right] \\
&= E\left[e^{tX_J}e^{tX_K}e^{tX_L}\right] \\
&= E\left[e^{tX_J}\right]E\left[e^{tX_K}\right]E\left[e^{tX_L}\right] \\
&= M_J(t)M_K(t)M_L(t) \\
&= (1-2t)^{-3}(1-2t)^{-2.5}(1-2t)^{-4.5} \\
&= (1-2t)^{-10}.
\end{aligned}$$

So that

$$\begin{aligned}
M_X'(t) &= (-10)(-2)(1-2t)^{-11} = 20(1-2t)^{-11}, \\
M_X''(t) &= 440(1-2t)^{-12}, \\
M_X'''(t) &= 10,560(1-2t)^{-13}.
\end{aligned}$$

Therefore, by Remark 2.2.7(4),

$$E[X^3] = M_X'''(0) = 10,560$$

Question 2.2.7. ([4], No. 60) A recent study indicates that the annual cost of maintaining and repairing a car in a town in Ontario averages 200 with a variance of 260.

If a tex of 20% is introduced on all items associated with the maintenance and repair of cars(i.e., everything is made 20% more expensive), what will be the variance of the annual cost of maintaining and repairing a car?

SOLUTION. Let

$$\begin{aligned} X = &\text{the annual cost of maintaining and repairing a car} \\ &\text{before the 20\% tax,} \\ Y = &\text{the annual cost of maintaining and repairing a car} \\ &\text{after the 20\% tax,.} \end{aligned}$$

The question is to find $Var(Y)$. We are given that

$$E[X] = 200, \quad Var(X) = 260$$

and

$$Y = (1 + 0.2)X = 1.2X.$$

So we have

$$Var(Y) = Var(1.2X) = (1.2)^2 Var(X) = (1.44)(260) = 374.$$

Question 2.2.8. ([4], No. 62) A random variable X has the cumulative distribution function

$$F_X(x) = \begin{cases} 0 & \text{for} \quad x < 1 \\ \frac{x^2-2x+2}{2} & \text{for} \quad 1 \le x < 2 \\ 1 & \text{for} \quad x \ge 2. \end{cases}$$

Calculate the variance of X.

SOLUTION. Note that if the cumulative distribution function $F_X(x)$ is discontinuous at some point $x = a$, then X has pmf

$$p_X(a) = \textit{height of the jump.}$$

Our distribution function is discontinuous at only one point $x = 1$. Thus X has the pmf

$$p_X(1) = \frac{1}{2} \quad \text{at} \quad x = 1,$$

and the pdf

$$f_X(x) = F'_X(x) = \begin{cases} x - 1 & \text{for} \quad 1 < x < 2 \\ 0 & \text{for} \quad x < 1, \ x \geq 2. \end{cases}$$

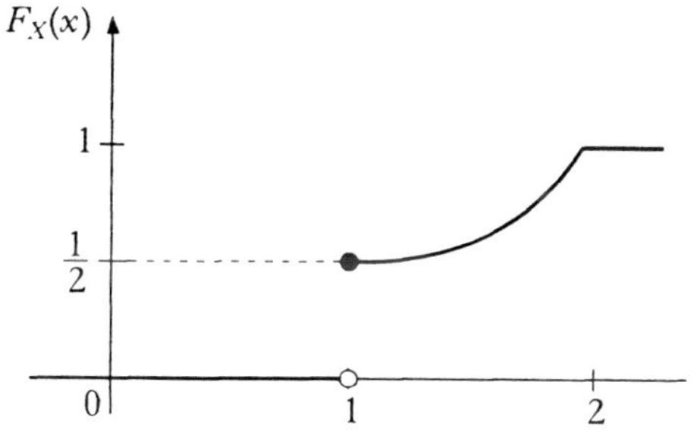

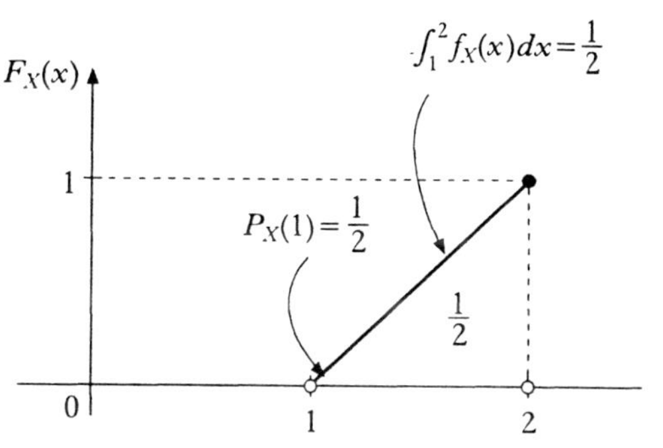

Then by Remark 2.2.5, we have

$$\begin{aligned} E[X] &= 1p_X(1) + \int_1^2 x(x-1)dx \\ &= \frac{1}{2} + \left[\frac{x^3}{3} - \frac{x^2}{2}\right]_1^2 = \frac{3}{4}, \end{aligned}$$

$$\begin{aligned} E[X^2] &= 1^2 p_X(1) + \int_1^2 x^2(x-1)dx \\ &= \frac{1}{2} + \left[\frac{x^4}{4} - \frac{x^3}{3}\right]_1^2 = \frac{23}{12} \end{aligned}$$

and so

$$\begin{aligned} Var(X) &= E[X^2] - E[X]^2 \\ &= \frac{23}{12} - \left(\frac{3}{4}\right)^2 = \frac{5}{36}. \end{aligned}$$

Question 2.2.9. ([4], No. 63) The warranty on a machine specifies that it will be replaced at failure or age 4, whichever occurs first. The machine's age at failure, X, has density function

$$f_X(x) = \begin{cases} \frac{1}{5} & \text{for} \quad 0 < x < 5, \\ 0 & \text{otherwise.} \end{cases}$$

Let Y be the age of the machine at the time of replacement.

Determine the variance of Y.

SOLUTION. Since

$$Y = g(X) = \begin{cases} X & \text{if} \quad 0 \leq X \leq 4, \\ 4 & \text{if} \quad 4 < X, \end{cases}$$

we have

$$\begin{aligned} E[Y] = E[g(X)] &= \int_0^5 g(x) f_X(x) dx \\ &= \int_0^4 x \frac{1}{5} dx + \int_4^5 4 \frac{1}{5} dx \\ &= \frac{12}{5}, \end{aligned}$$

$$\begin{aligned} E[Y^2] = E[g(X)^2] &= \int_0^5 g(x)^2 f_X(x) dx \\ &= \int_0^4 x^2 \frac{1}{5} dx + \int_4^5 4^2 \frac{1}{5} dx \\ &= \frac{112}{15} \end{aligned}$$

and so

$$\begin{aligned} Var(Y) &= E[Y^2] - E[Y]^2 \\ &= \frac{112}{15} - \left(\frac{12}{5}\right)^2 = 1.71. \end{aligned}$$

Question 2.2.10. ([4], No. 64) A probability distribution of the claim sizes for an auto insurance policy is given in the table below:

Claim Size	Probability
20	0.15
30	0.10
40	0.05
50	0.20
60	0.10
70	0.10
80	0.30

What percentage of the claims are within one standard deviation of the mean claim size?

SOLUTION. Let X be the claim size. One standard deviation of the mean claim size is an interval

$$M = \left[E[X] - 1 \cdot \sqrt{Var(X)}, E[X] + 1 \cdot \sqrt{Var(X)}\right].$$

Since

$$\begin{aligned} E[X] &= \sum_{k=1}^{7} k p_X(k) \\ &= (20)(0.15) + (30)(0.10) + (40)(0.05) + (50)(0.20) \\ &\quad + (60)(0.10) + (70)(0.10) + (80)(0.30) \\ &= 55, \end{aligned}$$

$$\begin{aligned}
E[X^2] &= \sum_{k=1}^{7} k^2 p_X(k) \\
&= (20)^2(0.15) + (30)^2(0.10) + (40)^2(0.05) + (50)^2(0.20) \\
&\quad + (60)^2(0.10) + (70)^2(0.10) + (80)^2(0.30) \\
&= 3,500
\end{aligned}$$

and

$$Var(X) = 3,500 - 55^2 = 475,$$

we have

$$M = [55 - \sqrt{475}, 55 + \sqrt{475}] = [33.21, 76.79].$$

Since $40, 50, 60, 70 \in [33.21, 76.79]$, the percentage of the claims are within $[33.21, 76.79]$ is

$$0.05 + 0.20 + 0.10 + 0.10 = 0.45.$$

Question 2.2.11. ([4], No. 66) A company agrees to accept the highest of four sealed bids on a property. The four bids are regarded as four independent random variables with common cumulative distribution function

$$F(x) = \frac{1}{2}(1 + \sin \pi x) \quad \text{for} \quad \frac{3}{2} \le x \le \frac{5}{2}.$$

Which of the following represents the expected value of the accepted bid?

(A) $\pi \int_{3/2}^{5/2} x \cos \pi x dx$

(B) $\frac{1}{16} \int_{3/2}^{5/2} (1 + \sin \pi x)^4 dx$

(C) $\frac{1}{16} \int_{3/2}^{5/2} x(1 + \sin \pi x)^4 dx$

(D) $\frac{1}{4}\pi \int_{3/2}^{5/2} \cos \pi x(1 + \sin \pi x)^3 dx$

(E) $\frac{1}{4}\pi \int_{3/2}^{5/2} x \cos \pi x(1 + \sin \pi x)^3 dx$

SOLUTION. Let X_1, X_2, X_3, X_4 be four independent bids with common distribution function $F(x)$ and

$$Y = \max(X_1, X_2, X_3, X_4).$$

The question is to find $E[Y]$. To do this, let us determine the distribution function $F_Y(y)$. Since X_1, X_2, X_3, X_4 are independent, we see

$$\begin{aligned} F_Y(y) &= P(Y \leq y) \\ &= P(X_1 \leq y, X_2 \leq y, X_3 \leq y, X_4 \leq y) \\ &= P(X_1 \leq y)P(X_2 \leq y)P(X_3 \leq y)P(X_4 \leq y) \\ &= F(y)^4 \\ &= \left(\frac{1}{2}\right)^4 (1 + \sin \pi y)^4 \end{aligned}$$

and so

$$\begin{aligned} f_Y(y) = F_Y'(y) &= \left(\frac{1}{2}\right)^4 4(1 + \sin \pi y)^3 (\cos \pi x)\pi \\ &= \frac{\pi}{4} \cos \pi y (1 + \sin \pi y)^3. \end{aligned}$$

Therefore we have

$$\begin{aligned} E[Y] &= \int_{3/2}^{5/2} y f_Y(y) dy \\ &= \frac{\pi}{4} \int_{3/2}^{5/2} y \cos \pi y (1 + \sin \pi y)^3 dy \end{aligned}$$

and so the answer is (E).

Question 2.2.12. ([4], No. 69) The time to failure of a component in an electronic device has an exponential distribution with a median of four hours.

Calculate the probability that the component will work without failing for at least five hours.

SOLUTION. Let T be the time to failure of a component in an electronic device. We are given that $T \sim E(\lambda)$, i.e.,

$$f_T(t) = \begin{cases} \lambda e^{-\lambda t} & \text{for} \quad t > 0, \\ 0 & \text{otherwise} \end{cases}$$

and that

$$P(T \leq 4) = 0.5.$$

The question is to calculate $P(T \geq 5)$. By assumption,

$$\begin{aligned} P(T \leq 4) &= \int_0^4 f_T(t)dt \\ &= \int_0^4 \lambda e^{-\lambda t} dt \\ &= \left[-e^{-\lambda t}\right]_0^4 = 1 - e^{-4\lambda}. \end{aligned}$$

Thus

$$0.5 = 1 - e^{-4\lambda}$$

and so

$$\lambda = \frac{1}{4} \ln 2.$$

Therefore

$$\begin{aligned} P(T \geq 5) &= 1 - P(T < 5) \\ &= 1 - \int_0^5 \frac{1}{4} \ln 2 \exp\left\{-\left(\frac{1}{4} \ln 2\right) t\right\} dt \\ &= 1 - \left[-\exp\left\{-\left(\frac{1}{4} \ln 2\right) t\right\}\right]_0^5 \\ &= \exp\left\{-\frac{5}{4} \ln 2\right\} \\ &= \frac{1}{2^{5/4}} = 0.42. \end{aligned}$$

Question 2.2.13. ([4], No. 76) Claim amounts for wind damage to insured homes are independent random variables with common density function

$$f(x) = \begin{cases} \frac{3}{x^4} & \text{for} \quad x > 1, \\ 0 & \text{otherwise,} \end{cases}$$

where x is the amount of a claim in thousands. Suppose 3 such claims will be made.

What is the expected value of the largest of the three claims?

SOLUTION. Let X_1, X_2, X_3 be such three independent claims and define

$$Y = \max(X_1, X_2, X_3).$$

The question is to find $E[X]$. Let $F(x)$ be the common distribution function of X_1, X_2, X_3. Then

$$F(x) = \int_1^x f(x)dx = \int_1^x \frac{3}{x^4}dx = 1 - \frac{1}{x^3}.$$

To find $E[Y]$, we need to determine the pdf $f_Y(y)$ of Y. Since

$$\begin{aligned} F_Y(y) &= P(Y \leq y) = P(X_1 \leq y, X_2 \leq y, X_3 \leq y) \\ &= P(X_1 \leq y)P(X_2 \leq y)P(X_3 \leq y) \\ &= F(y)^3 = \left(1 - \frac{1}{y^3}\right)^3, \end{aligned}$$

we get

$$f_Y(y) = F_Y'(y) = 3\left(1 - \frac{1}{y^3}\right)^2 \frac{3}{y^4} = \frac{9}{y^4}\left(1 - \frac{1}{y^3}\right)^2$$

for $y > 1$. So we have

$$
\begin{aligned}
E[Y] &= \int_1^\infty y f_Y(y)dy \\
&= \int_1^\infty y \frac{9}{y^4}\left(1 - \frac{1}{y^3}\right)^2 dy \\
&= \int_1^\infty \frac{9}{y^3}\left(1 - \frac{1}{y^3}\right)^2 dy \\
&= \int_1^\infty \left(\frac{9}{y^3} - \frac{18}{y^6} + \frac{9}{y^9}\right) dy \\
&= \left[-\frac{9}{2y^2} + \frac{18}{5y^5} - \frac{9}{8y^8}\right]_1^\infty \\
&= 9\left(\frac{1}{2} - \frac{2}{5} + \frac{1}{8}\right) \\
&= 2.025(\text{in thousand}) = 2025.
\end{aligned}
$$

Question 2.2.14. ([4], No. 96) A tour operator has a bus that can accommodate 20 tourists. The operator knows that tourists may not show up, so he sells 21 tickets. The probability that an individual tourist will not show up is 0.02, independent of all other tourists.
Each ticket costs is 50, and is non-refundable if a tourist fail to show up. If a tourist shows up and a seat is not available, the tour operator has to pay 100(=ticket cost+50 penalty) to the tourist.

What is the expected revenue of the tour operator?

SOLUTION. Let X be the revenue of the tour operator. The question is to find $E[X]$. Since the probability that all 21 tourists will show up is $(0.98)^{21} = 0.65$, the probability that at least one tourist will not show up is

$1 - 0.65 = 0.35$. So X is given by

$$X = \begin{cases} 950 & \text{if all 21 tourists show up} \\ 1050 & \text{if at least one tourist do not show up,} \end{cases}$$

$$= \begin{cases} 950 & \text{with probability } 0.65 \\ 1050 & \text{with probability } 0.35. \end{cases}$$

So we have

$$\begin{aligned} E[Y] &= (950)(0.65) + (1050)(0.35) \\ &= 985. \end{aligned}$$

Question 2.2.15. ([4], No. 98) Let X_1, X_2, X_3 be a random sample from a distribution with probability function

$$p(x) = \begin{cases} \frac{1}{3} & \text{for} \quad x = 0, \\ \frac{2}{3} & \text{for} \quad x = 1, \\ 0 & \text{otherwise.} \end{cases}$$

Determine the moment generating function $M_Y(t)$ of $Y = X_1X_2X_3$.

SOLUTION. We can see that Y has the values as follows;

$$Y = \begin{cases} 1 & \text{if} \quad X_1 = X_2 = X_3 = 1 \\ 0 & \text{otherwise,} \end{cases}$$

$$= \begin{cases} 1 & \text{w.p.} \quad P(X_1 = 1, X_2 = 1, X_3 = 1) = \left(\frac{2}{3}\right)^3 = \frac{8}{27} \\ 0 & \text{w.p.} \quad 1 - \frac{8}{27} = \frac{19}{27}. \end{cases}$$

So it holds

$$p_Y(y) = P(Y = y) = \begin{cases} \frac{19}{27} & \text{for} \quad y = 0, \\ \frac{8}{27} & \text{for} \quad y = 1, \\ 0 & \text{otherwise.} \end{cases}$$

Therefore

$$\begin{aligned} M_Y(t) = E\left[e^{tY}\right] &= e^{(t)(0)}\frac{19}{27} + e^{(t)(1)}\frac{8}{27} \\ &= \frac{19}{27} + e^t\frac{8}{27}. \end{aligned}$$

Question 2.2.16. ([**4**], No. 100) A car dealership sells 0, 1, or 2 luxury cars on any day. When selling a car, the dealer also tries to persuade the customer to buy an extended warranty for the car. Let X denote the number of luxury cars sold in a given day, and let Y denote the number of extended warranties sold.

$$\begin{aligned} P(X = 0, Y = 0) &= \frac{1}{6} \\ P(X = 1, Y = 0) &= \frac{1}{12} \\ P(X = 1, Y = 1) &= \frac{1}{6} \\ P(X = 2, Y = 0) &= \frac{1}{12} \\ P(X = 2, Y = 1) &= \frac{1}{3} \\ P(X = 2, Y = 2) &= \frac{1}{6} \end{aligned}$$

What is the variance of X?

SOLUTION. To find $Var(X)$, we need to determine $P(X = k)$, $k = 1, 2, 3$. Clearly $P(X = 0) = P(X = 0, Y = 0) = 1/6$. For the case $X = 1$, we have

$$\begin{aligned} P(X = 1) &= P(\{X = 1\} \cap S) \\ &= P(\{X = 1\} \cap (\{Y = 0\} \cup \{Y = 1\})) \\ &= P((\{X = 1\} \cap \{Y = 0\}) \cup (\{X = 1\} \cap \{Y = 1\}))) \\ &= P(\{X = 1\} \cap \{Y = 0\}) + P(\{X = 1\} \cap \{Y = 1\}) \\ &= P(X = 1, Y = 0) + P(X = 1, Y = 1) \\ &= \frac{1}{12} + \frac{1}{6} = \frac{3}{12}. \end{aligned}$$

Similarly

$$\begin{aligned} P(X=2) &= P(X=2, Y=0) + P(X=2, Y=1) + P(X=2, Y=2) \\ &= \frac{1}{12} + \frac{1}{3} + \frac{1}{6} = \frac{7}{12}. \end{aligned}$$

Then

$$\begin{aligned} E[X] &= (0)P(X=0) + (1)P(X=1) + (2)P(X=2) \\ &= (0)\left(\frac{1}{6}\right) + (1)\left(\frac{3}{12}\right) + (2)\left(\frac{7}{12}\right) \\ &= \frac{17}{12} \end{aligned}$$

and

$$\begin{aligned} E[X^2] &= (0)^2P(X=0) + (1)^2P(X=1) + (2)^2P(X=2) \\ &= (0)^2\left(\frac{1}{6}\right) + (1)^2\left(\frac{3}{12}\right) + (2)^2\left(\frac{7}{12}\right) \\ &= \frac{31}{12}. \end{aligned}$$

Therefore

$$Var(X) = \frac{31}{12} - \left(\frac{17}{12}\right)^2 = 0.58.$$

2.3. Some special discrete distributions

Question 2.3.1. ([4], No. 28) A hospital receives 1/5 of its flu vaccine shipments from Company X and the remainder of its shipments from other companies. Each shipment contains a very large number of vaccine vials. For Company X' shipments, 10% of the vials are ineffective. For every other company, 2% of the vials are ineffective. The hospital tests 30 randomly selected vials from a shipment and find one vial is ineffective.

What is the probability that this shipment came from Company X?

SOLUTION. Let

C : event that a shipment came from Company X

I : event that one of 30 vaccine vials tested is ineffective.

The question is to calculate $P(C|I)$. We use Bayes' rule

$$P(C|I) = \frac{P(I|C)P(C)}{P(I|C)P(C) + P(I|C^c)P(C^c)}.$$

We are given that $P(C) = 0.2$ and so $P(C^c) = 0.8$. And we can get

$$P(I|C) = \binom{30}{1}(0.1)^1(1-0.1)^{29} = 0.141$$

and similarly

$$P(I|C^c) = \binom{30}{1}(0.02)^1(1-0.02)^{29} = 0.334.$$

Hence we have

$$P(C|I) = \frac{(0.141)(0.2)}{(0.141)(0.2) + (0.334)(0.8)} = 0.096.$$

Question 2.3.2. ([4], No. 30) An actuary has discovered that policyholders are three times as likely to file two claims as to file four claims.

If the number of claims filed has a Poisson distribution, what is the variance of the number of claims filed?

SOLUTION. Let N be the number of claims filed. The question is to find $Var(N)$. Since $N \sim P(\lambda)$ for some λ, i.e,

$$p_N(n) = \frac{e^{-\lambda}\lambda^n}{n!},$$

we know that, by Theorem 2.3.5,

$$Var(N) = E[N] = \lambda.$$

So it is sufficient to determine λ. Since we are given that

$$P(N = 2) = 3P(N = 4),$$

it holds

$$\frac{e^{-\lambda}\lambda^2}{2!} = 3\frac{e^{-\lambda}\lambda^4}{4!}$$

and so $\lambda = 2$. Therefore, $Var(N) = 2$.

Question 2.3.3. ([4], No. 32) A large pool of adults earning their driver's license includes 50% low-risk drivers, 30% moderate-risk drivers, and 20% high-risk drivers. Because these drivers have no prior driving record, an insurance company considers each driver to be randomly selected from the pool.
This month, the insurance company writes 4 new policies for adults earning their first driver's license.

What is the probability that these 4 will contain at least two more high-risk than low-risk drivers?

SOLUTION. Let

$$\begin{aligned} X &= \text{number of low-risk drivers insured} \\ Y &= \text{number of moderate-risk drivers insured} \\ Z &= \text{number of high-risk drivers insured.} \end{aligned}$$

The question is to calculate $P(Z \geq X + 2)$. We are given that

$$(X, Y, Z) \sim M(4; 0.5, 0.3, 0.2)$$

. Then, by (2.3.7), we have

$$\begin{aligned} & P(Z \geq X + 2) \\ &= P(X = 0, Y = 0, Z = 4) + P(X = 0, Y = 1, Z = 3) \\ &\quad + P(X = 0, Y = 2, Z = 2) + P(X = 1, Y = 0, Z = 3) \\ &= \frac{4!}{0!0!4!}(0.5)^0(0.3)^0(0.2)^4 + \frac{4!}{0!1!3!}(0.5)^0(0.3)^1(0.2)^3 \\ &\quad + \frac{4!}{0!2!2!}(0.5)^0(0.3)^2(0.2)^2 + \frac{4!}{1!0!3!}(0.5)^1(0.3)^0(0.2)^3 \\ &= 0.0488. \end{aligned}$$

Question 2.3.4. ([4], No. 39) A company prices its hurricane insurance using the following assumption:

(i) In any calendar year, there can be at most one hurricane.

(ii) In any calendar year, the probability of a hurricane is 0.05.

(iii) The number of hurricane in any calendar year is independent of the number of hurricanes in any other calendar year.

Using the company's assumptions, calculate the probability that there are fewer than 3 hurricane in 20-year period.

SOLUTION. Let X be the number of hurricanes in a 20-year period. The question is to calculate $P(X < 3)$. By the company's assumptions, we are given that $X \sim B(20, 0.05)$. Thus

$$\begin{aligned}
&P(X < 3)\\
&= P(X = 0) + P(X = 1) + P(X = 2)\\
&= \binom{20}{0}(0.05)^0(0.95)^{20} + \binom{20}{1}(0.05)^1(0.95)^{19} + \binom{20}{2}(0.05)^2(0.95)^{18}\\
&= (1)(0.05)^0(0.95)^{20} + (20)(0.05)(0.95)^{19} + (190)(0.05)^2(0.95)^{18}\\
&= 0.925
\end{aligned}$$

Question 2.3.5. ([4], No. 41) A study is being conducted in which the health of two independent groups of ten policyholders is being monitored over a one-year period of time. Individual participants in the study drop out before the end of the study with probability 0.2 (independently of other participants).

What is the probability that at least 9 participants complete the study in one of the two groups, but not in both groups?

SOLUTION. Let

$$X = \text{number of group I participants that complete the study}$$

$$Y = \text{number of group II participants that complete the study.}$$

The question is to calculate $P(\{X \geq 9, Y < 9\} \cup \{X < 9, Y \geq 9\})$. We are given that $X, Y \sim B(10, 0.8)$ and they are independent. So we have

$$\begin{aligned}
&P(\{X \geq 9, Y < 9\} \cup \{X < 9, Y \geq 9\}) \\
&= P(X \geq 9, Y < 9) + P(X < 9, Y \geq 9) \\
&= 2P(X \geq 9, Y < 9) \\
&= 2P(X \geq 9)P(Y < 9) \\
&= 2P(X \geq 9)P(X < 9) \\
&= 2P(X \geq 9)(1 - P(X \geq 9)).
\end{aligned}$$

Since $X \sim B(10, 0.8)$, by (2.3.3),

$$\begin{aligned}
P(X \geq 9) &= P(X = 9) + P(X = 10) \\
&= \binom{10}{9}(0.8)^9(0.2)^1 + \binom{10}{10}(0.8)^{10}(0.2)^0 \\
&= 0.376.
\end{aligned}$$

Therefore,

$$\begin{aligned}
&P(\{X \geq 9, Y < 9\} \cup \{X < 9, Y \geq 9\}) \\
&= (2)(0.376)(1 - 0.376) \\
&= 0.469.
\end{aligned}$$

Question 2.3.6. ([4], No. 43) A company takes out an insurance policy to cover accidents that occur at its manufacturing plant. The probability that one or more accidents will occur during any given month is 3/5. The number of accidents that occur in any given month is independent of the number of accidents that occur in all other months.

Calculate the probability that there will be at least four months in which no accidents occur before the fourth month in which at least one accident occurs.

SOLUTION. Let us regard a month with one or more accidents as a success, and let X be the number of failures before the 4th success. The question is to calculate $P(X \geq 4)$. We are given that

$$X \sim NB\left(4, \frac{2}{5}\right).$$

By (2.3.4), the pmf of X is given by

$$p_X(k) = P(X = k) = \binom{k+4-1}{k}\left(\frac{3}{5}\right)^4\left(\frac{2}{5}\right)^k.$$

Thus we have

$$\begin{aligned}
P(X \geq 4) &= 1 - P(X \leq 3) \\
&= 1 - \{P(X=0) + P(X=1) + P(X=2) + P(X=3)\} \\
&= 1 - \left\{\binom{3}{0}\left(\frac{3}{5}\right)^4\left(\frac{2}{5}\right)^0 + \binom{4}{1}\left(\frac{3}{5}\right)^4\left(\frac{2}{5}\right)^1\right. \\
&\qquad \left. + \binom{5}{2}\left(\frac{3}{5}\right)^4\left(\frac{2}{5}\right)^2 + \binom{6}{3}\left(\frac{3}{5}\right)^4\left(\frac{2}{5}\right)^3\right\} \\
&= 0.2898.
\end{aligned}$$

Question 2.3.7. ([4], No. 50) A company buys a policy to insure its revenue in the event of major snowstorms that shut down business. The policy pays nothing for the first snowstorm of the year and 10,000 for each one thereafter, until the end of the year. The number of major snowstorms per year that shut down business is assumed to have a Poisson distribution with mean 1.5.

What is the expected amount paid to the company under this policy during a one-year period?

SOLUTION. Let

N = number of major snowstorms per year that shut down business

X = claim amount paid to the company under this policy.

The question is to find $E[X]$. We are given that $N \sim P(1.5)$, i.e.,

$$p_N(n) = \frac{(1.5)^n e^{-1.5}}{n!}$$

and

$$X = g(N) = \begin{cases} 0 & \text{if} \quad N = 0, \\ 10,000(N-1) & \text{if} \quad N = 1, 2, 3, \cdots. \end{cases}$$

Then

$$\begin{aligned} E[X] = E[g(N)] &= \sum_{n=0}^{\infty} g(n) p_N(n) \\ &= \sum_{n=1}^{\infty} 10,000(n-1) \frac{(1.5)^n e^{-1.5}}{n!} \\ &= 10,000e^{-1.5} + \sum_{n=0}^{\infty} 10,000(n-1) \frac{(1.5)^n e^{-1.5}}{n!} \\ &= 10,000e^{-1.5} + E[10,000(N-1)] \\ &= 10,000e^{-1.5} + 10,000E[N] - 10,000. \end{aligned}$$

Since $N \sim P(1.5)$, by Theorem 2.3.5, $E[N] = 1.5$. Therefore we get

$$E[X] = 10,000e^{-1.5} + (10,000)(1.5) - 10,000 = 7,231.$$

Question 2.3.8. ([4], No. 67) A baseball team has scheduled its opening game for April 1. If it rains on April 1, the game is postponed and will be played on the next day that it does not rain. The team purchases insurance against rain. The policy will pay 1,000 for each day, up to 2 days, that the opening game is postponed.
The insurance company determines that the number of consecutive days of rain beginning on April 1 is a Poisson random variable with mean 0.6.

What is the standard deviation of the amount the insurance company will have to pay?

SOLUTION. Let

X = number of consecutive days of rain beginning on April 1

Y = amount the insurance company will have to pay.

The question is to find $\sqrt{Var(Y)}$. We are given that $X \sim P(0.6)$, i.e.,

$$p_X(k) = \frac{(0.6)^k e^{-0.6}}{k!}$$

and

$$Y = g(X) = \begin{cases} 1,000X & \text{if} \quad X = 1, 2, \\ 2,000 & \text{if} \quad X = 3, 4, \cdots . \end{cases}$$

Then we have

$$\begin{aligned}
E[Y] = E[g(X)] &= \sum_{k=0}^{\infty} g(k) p_X(k) \\
&= (1,000)(1)\frac{e^{-0.6}(0.6)^1}{1!} + (1,000)(2)\frac{e^{-0.6}(0.6)^2}{2!} \\
&\quad \left\{ +2,000\frac{e^{-0.6}(0.6)^3}{3!} + 2,000\frac{e^{-0.6}(0.6)^4}{4!} + \cdots \right\} \\
&= 1,000\frac{e^{-0.6}(0.6)^1}{1!} + 2,000 \left\{ \sum_{k=2}^{\infty} \frac{e^{-0.6}(0.6)^k}{k!} \right\} \\
&= 1,000\frac{e^{-0.6}(0.6)^1}{1!} + 2,000 \left\{ \sum_{k=0}^{\infty} \frac{e^{-0.6}(0.6)^k}{k!} - e^{-0.6} - 0.6e^{-0.6} \right\} \\
&= 1,000\frac{e^{-0.6}(0.6)^1}{1!} + 2,000 \left\{ 1 - e^{-0.6} - 0.6e^{-0.6} \right\} \\
&= 573
\end{aligned}$$

and similarly

$$\begin{aligned}
E[Y^2] = E[g(X)^2] &= \sum_{k=0}^{\infty} g(k)^2 p_X(k) \\
&= (1,000)^2\frac{e^{-0.6}(0.6)^1}{1!} + (2,000)^2 \left\{ 1 - e^{-0.6} - 0.6e^{-0.6} \right\} \\
&= 816,893.
\end{aligned}$$

Therefore $\sqrt{Var(Y)} = \sqrt{816,893 - 573^2} = \sqrt{488,564} = 699.$

2.4. Some special continuous distributions

Question 2.4.1. ([4], No. 18) Two instruments are used to measure the height, h, of a tower. The error made by the less accurate instrument is normally distributed with mean 0 and standard deviation $0.0056h$. The error made by more accurate instrument is normally distributed with mean 0 and standard deviation $0.0044h$.

Assuming the measurements are independent random variables, what is the probability that their average value is within $0.005h$ of the height of the tower?

SOLUTION. Let X_1 and X_2 be the measurement error made by the less and more accurate instruments, respectively. And let

$$Y = \frac{X_1 + X_2}{2}.$$

The question is to find $P(|Y| \leq 0.005h)$. We are given that

$$X_1 \sim N\left(0, (0.0056h)^2\right) \quad \text{and} \quad X_2 \sim N\left(0, (0.0044h)^2\right).$$

Then

$$E[Y] = \frac{1}{2}(E[X_1] + E[X_2]) = 0$$

and

$$\begin{aligned} Var(Y) &= \frac{1}{4}\{Var(X_1) + Var(X_2)\} \\ &= \frac{1}{4}\{(0.0056h)^2 + (0.0044h)^2\} \\ &= (0.0028h)^2 + (0.0022h)^2 \\ &= (0.00356h)^2. \end{aligned}$$

Using Remark 2.4.10, we see

$$Y \sim N\left(0, (0.00356h)^2\right).$$

Thus we have, from the table for standard normal distribution,

$$\begin{aligned}
P(|Y| \leq 0.005h) &= P(-0.005h \leq Y \leq 0.005h) \\
&= 2P(Y \leq 0.005h) - 1 \\
&= 2P\left(\frac{Y}{0.00356h} \leq \frac{0.005h}{0.00356h}\right) - 1 \\
&= 2P\left(\frac{Y}{0.00356h} \leq 1.4\right) - 1 \\
&= 2\Phi(1.4) - 1 \\
&= (2)(0.9192) - 1 = 0.84.
\end{aligned}$$

Question 2.4.2. ([4], No. 29) The number of days that elapse between the beginning of a calender year and the moment a high-risk driver is involved in an accident is exponential distributed. An insurance company expects that 30% of high-risk drivers will be involved in an accident during the first 50 days of a calender year.

What portion of high-risk drivers are expected to be involved in an accident during the first 80 days of a calender year?

SOLUTION. Let T be the number of days that elapse between the beginning of a calender year and the moment a high-risk driver is involved in an accident. The question is to find $P(T \leq 80)$. We are given that

(i) $T \sim E(\lambda)$, i.e., $f_T(t) = \lambda e^{-\lambda t}$, for some $\lambda > 0$,

(ii) $P(T \leq 50) = 0.3$.

Let us determine λ from (i)and (ii).

$$\begin{aligned}
0.3 = P(T \leq 50) &= \int_0^{50} f_T(t)dt \\
&= \int_0^{50} \lambda e^{-\lambda t} dt \\
&= \left[-e^{-\lambda t}\right]_{t=0}^{t=50} = 1 - e^{-50\lambda}.
\end{aligned}$$

From this we have

$$e^{-50\lambda} = 0.7 \quad \text{or} \quad -50\lambda = \ln(0.7)$$

and hence

$$\lambda = -\frac{1}{50}\ln(0.7).$$

Thus

$$\begin{aligned} P(T \le 80) &= \int_0^{80} \lambda e^{-\lambda t} dt = 1 - e^{-80\lambda} \\ &= 1 - e^{-\frac{80}{50}\ln(0.7)} = 1 - (0.7)^{-1.6} = 0.435. \end{aligned}$$

Question 2.4.3. ([4], No. 42) For Company A there is a 60% chance that no claim is made during the coming year. If one or more claims are made, total claim amount is normally distributed with mean 10,000 and standard deviation 2,000.
For Company B there is 70% chance that no claim is made during the coming year. If one or more claims are made, total claim amount is normally distributed with mean 9,000 and standard deviation 2,000.
Assume that the total claim amounts of two companies are independent.

What is the probability that, in the coming year, Company B's total claim amount will exceed Company A's total claim amounts?

SOLUTION. Let

M_A = event that Company A makes a claim

M_B = event that Company B makes a claim

X_A = Company A's total claim amount if claims are made

X_B = Company B's total claim amount if claims are made

The question is to find $p = P((M_A^c \cap M_B) \cup \{(M_A \cap M_B) \cap (X_A < X_B)\})$.
From the assumptions, we are given that

$$P(M_A^c) = 0.6 \quad P(M_A) = 0.4 \quad X_A \sim N(10,000, 2,000^2)$$

$$P(M_B^c) = 0.7 \quad P(M_B) = 0.3 \quad X_B \sim N(9,000, 2,000^2).$$

Since M_A and M_B are independent,

$$\begin{aligned} p &= P((M_A^c \cap M_B) \cup \{(M_A \cap M_B) \cap (X_A < X_B)\}) \\ &= P(M_A^c)P(M_B) + P(M_A)P(M_B)P(X_A < X_B) \\ &= (0.6)(0.3) + (0.4)(0.3)P(X_B - X_A > 0). \end{aligned}$$

We have to calculate $P(X_B - X_A > 0)$. Since X_A and X_B are independent,

$$E[X_B - X_A] = E[X_B] - E[X_A] = 9,000 - 10,000 = -1,000$$

and

$$\begin{aligned} Var(X_B - X_A) &= Var(X_B) + Var(-X_A)\} \\ &= 2,000^2 + (-1)^2(2,000)^2 \\ &= (2,000\sqrt{2})^2. \end{aligned}$$

Using Remark 2.4.10, we see

$$X_B - X_A \sim N\left(-1,000, (2,000\sqrt{2})^2\right).$$

Thus we have, from the table for standard normal distribution,

$$\begin{aligned} P(X_B - X_A > 0) &= 1 - P(X_B - X_A < 0) \\ &= 1 - P\left(\frac{X_B - X_A + 1,000}{2,000\sqrt{2}} \le \frac{1,000}{2,000\sqrt{2}}\right) \\ &= 1 - P\left(\frac{X_B - X_A + 1,000}{2,000\sqrt{2}} \le \frac{1}{2\sqrt{2}}\right) \\ &= 1 - \Phi\left(\frac{1}{2\sqrt{2}}\right) = 1 - \Phi(0.354) \\ &= 1 - 0.638 = 0.362. \end{aligned}$$

Therefore it holds

$$\begin{aligned} p &= (0.6)(0.3) + (0.4)(0.3)P(X_B - X_A > 0) \\ &= (0.6)(0.3) + (0.4)(0.3)(0.362) = 0.223. \end{aligned}$$

Question 2.4.4. ([4], No. 46) A device that continuously measures and records seismic activity is placed in a remote region. The time, T, to failure of this device is exponentially distributed with mean 3 years. Since the device will not be monitored during its first two years of service, the time to discovery of its failure is $X = \max(T, 2)$.

Determine $E[X]$.

SOLUTION. Since $T \sim E\left(\frac{1}{3}\right)$, its pdf is given by

$$f_T(t) = \begin{cases} \frac{1}{3}e^{-\frac{1}{3}t}, & t > 0 \\ 0, & \text{otherwise.} \end{cases}$$

Since

$$X = g(T) = \max(T, 2) = \begin{cases} 2, & \text{if} \quad 0 < T < 2 \\ T, & \text{if} \quad T \geq 2, \end{cases}$$

we get

$$\begin{aligned} E[X] = E[g(T)] &= \int_{-\infty}^{\infty} g(t) f_T(t) dt \\ &= \int_0^2 (2)\left(\frac{1}{3}\right) e^{-\frac{1}{3}t} dt + \int_2^{\infty} t \left(\frac{1}{3}\right) e^{-\frac{1}{3}t} dt \\ &= -2e^{-\frac{2}{3}} + 2 + \left[-te^{-\frac{t}{3}}\right]_2^{\infty} + \int_2^{\infty} e^{-\frac{1}{3}t} dt \\ &= 2 + 3e^{-\frac{2}{3}}. \end{aligned}$$

Question 2.4.5. ([4], No. 47) A piece of equipment is being insured against early failure. The time from purchase until failure of the equipment is exponentially distributed with mean 10 years. The insurance will pay an amount x if the equipment fails during the first year, and it will pay $0.5x$ if failure occurs during the second or third year. If failure occurs after the first three years, no payment will be made.

At what level must x be set if the expected payment made under this insurance is to be 1,000?

SOLUTION. Let T be the time from purchase until failure of the equipment and X claim amounts for T. We are given that $T \sim E\left(\frac{1}{3}\right)$, i.e.,

$$f_T(t) = \begin{cases} \frac{1}{10}e^{-\frac{1}{10}t}, & t > 0 \\ 0, & \text{otherwise,} \end{cases}$$

and

$$X = g(T) = \begin{cases} x, & \text{if } \ 0 \le T \le 1 \\ x/2, & \text{if } \ 1 < T \le 3 \\ 0, & \text{if } \ T > 3. \end{cases}$$

The question is to determine x such that $E[X] = 1,000$. Since $E[X] = 1,000$, it follows that

$$\begin{aligned} 1,000 = E[X] = E[g(T)] &= \int_{-\infty}^{\infty} g(t) f_T(t) dt \\ &= \int_0^1 (x)\left(\frac{1}{10}\right) e^{-\frac{1}{10}t} dt + \int_1^3 \left(\frac{x}{2}\right)\left(\frac{1}{10}\right) e^{-\frac{1}{10}t} dt \\ &= 0.1772x. \end{aligned}$$

So we get $x = 5644$.

Question 2.4.6. ([4], No. 70) An insurance company sells an auto insurance policy that covers losses incurred by a policyholder, subject to a deductible of 100. Losses incurred follow an exponential distribution with mean 300.

What is the 95th percentile of actual losses that exceed the deductible?

SOLUTION. Let X be actual losses incurred. We are given that $X \sim E\left(\frac{1}{300}\right)$. The question is to determine x such that

$$P(X \le x | X > 100) = 0.95.$$

By the definition of the conditional probability, it follows

$$\begin{aligned}
0.95 = P(X \le x | X > 100) &= \frac{P(100 < X \le x)}{P(X > 100)} \\
&= \frac{P(X \le x) - P(X \le 100)}{1 - P(X \le 100)} = \frac{F_X(x) - F_X(100)}{1 - F_X(100)} \\
&= \frac{(1 - e^{-x/300}) - (1 - e^{-100/300})}{1 - (1 - e^{-100/300})} \\
&= 1 - e^{1/3} e^{-x/300}.
\end{aligned}$$

So we have

$$e^{-x/300} = 0.05 e^{-1/3}$$

and hence

$$x = -300 \ln \left(0.05 e^{-1/3}\right) = 999.$$

Question 2.4.7. ([4], No. 72) An investment account earns an annual interest rate R that follows a uniform distribution on the interval (0.04,0.08). The value of a 10,000 initial investment in this account after one year is given by $V = 10,000 e^R$.

Determine the cumulative distribution function, $F_V(v)$, of V for values of v that satisfy $0 < F_V(v) < 1$.

SOLUTION. Since $R \sim U(0.04, 0.08)$, its pdf is given by

$$f_R(r) = \begin{cases} \frac{1}{0.08 - 0.04} = \frac{1}{0.04}, & t \in (0.04, 0.08) \\ 0, & \text{otherwise}, \end{cases}$$

Thus

$$
\begin{aligned}
F_V(v) &= P(V \le v) = P(10,000e^R \le v) \\
&= P\left(e^R \le \frac{v}{10,000}\right) = P\left[R \le \ln\left(\frac{v}{10,000}\right)\right] \\
&= P(R \le \ln v - \ln 10,000) \\
&= \int_{0.04}^{\ln v - \ln 10,000} \frac{1}{0.04} dr \\
&= \left[\frac{r}{0.04}\right]_{r=0.04}^{r=\ln v - \ln 10,000} \\
&= 25\left[\ln\left(\frac{v}{10,000}\right) - 0.04\right].
\end{aligned}
$$

Question 2.4.8. ([4], No. 73) An actuary models the lifetime of a device using the random variable $Y = 10X^{0.8}$, where X is an exponential random variable with mean 1 year.

Determine the probability density function $f_Y(y)$, for $y > 0$, of the random variable Y.

SOLUTION. Since $X \sim E(1)$, the distribution function $F_X(x)$ is given by

$$F_X(x) = 1 - e^{-x}.$$

So that

$$
\begin{aligned}
F_Y(y) &= P(Y \le y) = P(10X^{0.8} \le y) \\
&= P\left(X \le \left(\frac{y}{10}\right)^{1.25}\right) \\
&= F_X\left(\left(\frac{y}{10}\right)^{1.25}\right) = 1 - \exp\left\{-\left(\frac{y}{10}\right)^{1.25}\right\}.
\end{aligned}
$$

Therefore we have

$$
\begin{aligned}
f_Y(y) = F'_Y(y) &= (1.25)\left(\frac{1}{10}\right)\left(\frac{y}{10}\right)^{0.25} \exp\left\{-\left(\frac{y}{10}\right)^{1.25}\right\} \\
&= (0.125)\left(\frac{y}{10}\right)^{0.25} \exp\left\{-\left(\frac{y}{10}\right)^{1.25}\right\}.
\end{aligned}
$$

Question 2.4.9. ([4], No. 74) Let T denote the time in minutes for a customer service representative to respond to 10 telephone inquiries. T is uniformly distributed on the interval with endpoints 8 minutes and 12 minutes. Let R denote the average rate, in customers per minute, at which the representative responds to inquiries.

Determine the probability density function $f_R(r)$ of the random variable R on the interval (10/12, 10/8).

SOLUTION. We are given that $T \sim U(8,12)$, i.e., $f_T(t) = 1/4$ on (8,12) and $R = 10/T$. Thus

$$\begin{aligned} F_R(r) &= P(R \le r) = P\left(\frac{10}{T} \le r\right) = P\left(T \ge \frac{10}{r}\right) \\ &= 1 - P\left(T \le \frac{10}{r}\right) = 1 - F_T\left(\frac{10}{r}\right). \end{aligned}$$

Therefore we have

$$\begin{aligned} f_R(r) &= F_R'(r) = \left(1 - F_T\left(\frac{10}{r}\right)\right)' \\ &= -F_T'\left(\frac{10}{r}\right)\left(\frac{-10}{r^2}\right) = -f_T\left(\frac{10}{r}\right)\left(\frac{-10}{r^2}\right) \\ &= -\left(\frac{1}{4}\right)\left(\frac{-10}{r^2}\right) = \frac{5}{2r^2}. \end{aligned}$$

Question 2.4.10. ([4], No. 80) A charity receives 2,025 contributions. Contributions are assumed to be independent and identically distributed with mean 3,125 and standard deviation 250.

Calculate the approximate 90th percentile for the distribution of the total contributions received.

SOLUTION. Let X_k denote k-th contribution received, $k = 1, 2, \cdots, 2025$ and $S = \sum_{k=1}^{2025} X_k$. The question is to determine x such that $P(S \le x) = 0.9$. Since, for any $k = 1, 2, \cdots, 2025$, $E[X_k] = 3,125$ and $Var(X_k) = 250^2$, we have

$$E[S] = \sum_{k=1}^{2025} E[X_k] = 2,025 \times 3,125 = 6,328,125.$$

and

$$Var(S) = \sum_{k=1}^{2025} Var(X_k) = 2,025 \times 250^2 = 45^2 \times 250^2.$$

By the standardization, we see that

$$Z \equiv \frac{S - E[S]}{\sqrt{Var(S)}} \sim N(0, 1^2).$$

Thus

$$\begin{aligned} 0.9 = P(S \le x) = P\left(\frac{S - E[S]}{\sqrt{Var(S)}} \le \frac{x - E[S]}{\sqrt{Var(S)}}\right) \\ = P\left(\frac{S - E[S]}{\sqrt{Var(S)}} \le \frac{x - 6,328,125}{45 \times 250}\right) \\ = \Phi\left(\frac{x - 6,328,125}{45 \times 250}\right), \end{aligned}$$

where $\Phi(z)$ is the pdf of $Z \sim N(0, 1^2)$. By the standard normal table, we can see that

$$\frac{x - 6,328,125}{45 \times 250} = 1.282.$$

From this we have $x = 6,343,000$.

Question 2.4.11. ([4], No. 83) A company manufactures a brand of light bulb with a lifetime in months that is normally distributed with mean 3 and variance 1. A consumer buys a number of these bulbs with the intention of replacing them successively as they burn out. The light bulbs have independent lifetimes.

What is the smallest number of bulbs to be purchased so that succession of light bulbs produces light for at least 40 months with probability at least 0.9772?

SOLUTION. Let X_i be the lifetime of i-th light bulb for $i = 1, 2, \cdots$, and let

$$S_n = X_1 + X_2 + \cdots + X_n.$$

The question is to find the smallest number n such that $P(S_n > 40) \geq 0.9772$. Since $X_i \sim N(3, 1^2)$ for all $i = 1, 2, \cdots$, and these are independent, we know that

$$E[S_n] = \sum_{i=1}^{n} E[X_i] = 3n.$$

and

$$Var(S) = \sum_{i=1}^{n} Var(X_i) = n.$$

By Remark 2.4.10, we see $S_n \sim N(3n, n)$ and hence

$$\begin{aligned} 0.9772 &\leq P(S_n > 40) \\ &= P\left(\frac{S_n - 3n}{\sqrt{n}} > \frac{40 - 3n}{\sqrt{n}}\right) \\ &= 1 - P\left(\frac{S_n - 3n}{\sqrt{n}} \leq \frac{40 - 3n}{\sqrt{n}}\right) \\ &\approx 1 - \Phi\left(\frac{40 - 3n}{\sqrt{n}}\right) \end{aligned}$$

On the other hand, from the standard normal table, we see that

$$0.9772 = \Phi(2) = 1 - \Phi(-2).$$

Thus

$$1 - \Phi(-2) \leq 1 - \Phi\left(\frac{40 - 3n}{\sqrt{n}}\right),$$

that is,

$$\Phi\left(\frac{40 - 3n}{\sqrt{n}}\right) \leq \Phi(-2)$$

and hence

$$\frac{40 - 3n}{\sqrt{n}} \leq -2.$$

Rewriting this,

$$3n - 2\sqrt{n} - 40 \geq 0$$

or

$$(3\sqrt{n}+10)(\sqrt{n}-4) \geq 0.$$

From this we have $\sqrt{n}=4$ and so $n=16$.

Question 2.4.12. ([4], No. 103) In a small metropolitan area, annual losses due to storm, fire, and theft are assumed to be independent, exponential distributed random variables with respective means 1.0, 1.5, and 2.4.

Determine the probability that the maximum of these losses exceeds 3.

SOLUTION. Let

$$\begin{aligned} X &= \text{annual losses due to storm} \\ Y &= \text{annual losses due to fire} \\ Z &= \text{annual losses due to theft} \\ M &= \max(X, Y, Z). \end{aligned}$$

The question is to determine $P(M > 3)$. We are given that

$$\begin{aligned} X &\sim E\left(\frac{1}{1.0}\right), \text{ i.e., } F_X(x) = 1 - e^{-\frac{1}{1.0}x}, \\ Y &\sim E\left(\frac{1}{1.5}\right), \text{ i.e., } F_Y(y) = 1 - e^{-\frac{1}{1.5}y}, \\ Z &\sim E\left(\frac{1}{2.4}\right), \text{ i.e., } F_Z(z) = 1 - e^{-\frac{1}{2.4}z}. \end{aligned}$$

Thus it follows

$$\begin{aligned} P(M > 3) &= 1 - P(M \leq 3) \\ &= 1 - P(X \leq 3, Y \leq 3, Z \leq 3) \\ &= 1 - P(X \leq 3)P(Y \leq 3)P(Z \leq 3) \\ &= 1 - FX(3)F_Y(3)F_Z(3) \\ &= 1 - \left\{1 - e^{-\frac{3}{1.0}}\right\}\left\{1 - e^{-\frac{3}{1.5}}\right\}\left\{1 - e^{-\frac{3}{2.4}}\right\} \\ &= 0.414. \end{aligned}$$

CHAPTER 3

Joint and Conditional Distributions

3.1. Joint distributions

Question 3.1.1. ([4], No. 77) A device runs until either of two components fails, at which point the device stops running. The joint density function of the lifetimes of the two components, both measured in hours, is

$$f(x,y) = \begin{cases} \frac{x+y}{8}, & \text{for } \ 0 < x, y < 2 \\ 0, & \text{otherwise.} \end{cases}$$

What is the probability that the device fails during its first hour of operation?

SOLUTION. Let X, Y be the lifetimes of the two components. The question is to find $P(X < 1 \text{ or } Y < 1)$. Let D be the support of the joint probability density function $f(x,y)$, i.e., $D = \{(x,y)|0 < x, y < 2\}$. Then $P(X < 1 \text{ or } Y < 1) = P((X,Y) \in A)$, where $A = \{(x,y) \in D|x < 1 \text{ or } y < 1\}$. Since $A = \{(x,y)|0 < x < 1, 0 < y < 2\} \cup \{(x,y)|1 < x < 2, 0 < y <$

$1\} = A_1 \cup A_2$, we have

$$
\begin{aligned}
&P(X < 1 \text{ or } Y < 1)\\
&= P((X,Y) \in A_1 \cup A_2)\\
&= P((X,Y) \in A_1) + P((X,Y) \in A_2)\\
&= \int\int_{A_1} f(x,y)dxdy + \int\int_{A_2} f(x,y)dxdy\\
&= \int_0^2 \left[\int_0^1 \frac{1}{8}(x+y)dx\right] dy + \int_0^1 \left[\int_1^2 \frac{1}{8}(x+y)dx\right] dy\\
&= \int_0^2 \left[\frac{x^2+2xy}{16}\right]_0^1 dy + \int_0^1 \left[\frac{x^2+2xy}{16}\right]_1^2 dy\\
&= \int_0^2 \frac{1+2y}{16} dy + \int_0^1 \frac{3+2y}{16} dy\\
&= \frac{1}{16}[y+y^2]_0^2 + \frac{1}{16}[3y+y^2]_0^1\\
&= \frac{10}{16} = 0.625.
\end{aligned}
$$

Alternatively, since $A^c(= D \cap A^c) = \{(x,y) \in D | 1 \le x < 2, 1 \le y < 2\}$, we have

$$
\begin{aligned}
&P(X < 1 \text{ or } Y < 1)\\
&= 1 - P((X,Y) \in A^c)\\
&= 1 - \int\int_{A^c} f(x,y)dxdy\\
&= 1 - \int_1^2 \left[\int_1^2 \frac{1}{8}(x+y)dx\right] dy\\
&= 1 - \int_1^2 \left[\frac{x^2+2xy}{16}\right]_1^2 dy\\
&= 1 - \int_1^2 \frac{3+2y}{16} dy\\
&= 1 - \frac{1}{16}[3y+y^2]_1^2\\
&= \frac{10}{16} = 0.625.
\end{aligned}
$$

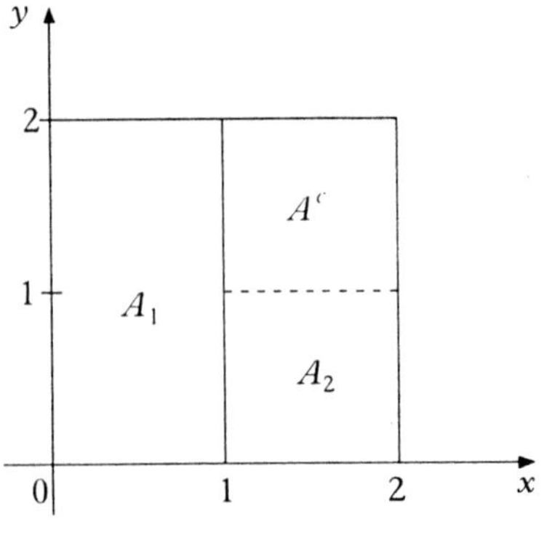

Question 3.1.2. ([4], No. 88) The waiting time for the first claim from a good driver and the waiting time for the first claim from a bad driver are independent and follow exponential distributions with means 6 years and 3 years, respectively.

What is the probability that the first claim from a good driver will be filed within 3 years and the first claim from a bad driver will be filed within 2 years?

SOLUTION. Let X and Y be the waiting times for the first claim from a good driver and a bad driver, respectively. The question is to find $P(X \leq 3, Y \leq 2)$. We are given that

$$X \sim E(1/6), \quad \text{i.e.,} \quad f_X(x) = \frac{1}{6}e^{-\frac{1}{6}x}, \quad x > 0,$$
$$Y \sim E(1/3), \quad \text{i.e.,} \quad f_Y(y) = \frac{1}{3}e^{-\frac{1}{3}y}, \quad y > 0.$$

Since X, Y are independent, we have

$$\begin{aligned} P(X \leq 3, Y \leq 2) &= P(X \leq 3)P(Y \leq 2) \\ &= \int_0^3 \frac{1}{6}e^{-\frac{1}{6}x}dx \int_0^2 \frac{1}{3}e^{-\frac{1}{3}y}dy \\ &= \left[-e^{-\frac{1}{6}x}\right]_0^3 \left[-e^{-\frac{1}{3}y}\right]_0^2 \\ &= \left(1 - e^{-\frac{1}{2}}\right)\left(-e^{-\frac{2}{3}}\right) \\ &= 1 - e^{-\frac{2}{3}} - e^{-\frac{1}{2}} + e^{-\frac{7}{6}}. \end{aligned}$$

Question 3.1.3. ([4], No. 89) The future lifetimes (in months) of two components of a machine have the following joint density function:

$$f(x, y) = \begin{cases} \frac{6}{125{,}000}(50 - x - y), & \text{for} \quad 0 < x < 50 - y < 50 \\ 0, & \text{otherwise,} \end{cases}$$

What is the probability that both components are still functioning 20 months from now?

(A) $\frac{6}{125{,}000}\int_0^{20}\int_0^{20}(50 - x - y)dydx$

(B) $\frac{6}{125{,}000}\int_{20}^{30}\int_{20}^{50-x}(50-x-y)dydx$
(C) $\frac{6}{125{,}000}\int_{20}^{30}\int_{20}^{50-x-y}(50-x-y)dydx$
(D) $\frac{6}{125{,}000}\int_{20}^{50}\int_{20}^{50-x}(50-x-y)dydx$
(E) $\frac{6}{125{,}000}\int_{20}^{50}\int_{20}^{50-x-y}(50-x-y)dydx$

SOLUTION. Let X, Y be the future lifetimes (in months) of two components of a machine. The question is to find $P(X > 20, Y > 20)$. And let D be the support of the joint pdf $f(x, y)$, i.e., $D = \{(x, y)|0 < x < 50, 0 < y < 50 - x\}$. Then $P(X > 20, Y > 20) = P((X, Y) \in A)$, where $A = \{(x, y) \in D|x > 20, y > 20\} = \{(x, y)|20 < x < 30, 20 < y < 50 - x\}$.

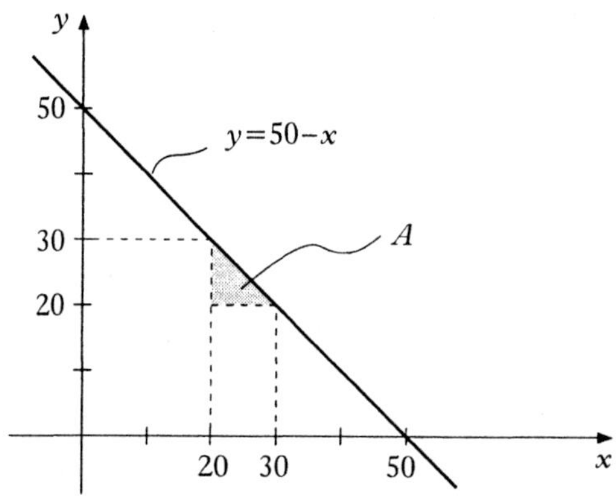

Hence we have

$$\begin{aligned} P(X > 20, Y > 20) &= P((X, Y) \in A) \\ &= \int\int_A f(x, y)dxdy \\ &= \int_{20}^{30}\left[\int_{20}^{50-x}\frac{6}{125{,}000}(50 - x - y)dy\right]dx. \end{aligned}$$

So the answer is (B).

Question 3.1.4. ([4], No. 90) An insurance company sells two types of auto insurance policies: Basic and Deluxe. The time until the next Basic Policy claim is an exponential random variable with mean two days.

The time until the next Deluxe Policy claim is an independent exponential random variable with mean three days.

What is the probability that the next claim will be a Deluxe Policy claim?

SOLUTION. Let

T_1 : time until the next Basic Policy claim

T_2 : time until the next Deluxe Policy claim.

The question is to find $P(T_2 < T_1)$. We are given that

$$X \sim E(1/2), \quad \text{i.e.,} \quad f_{T_1}(t_1) = \frac{1}{2}e^{-\frac{1}{2}t_1}, \quad t_1 > 0,$$
$$Y \sim E(1/3), \quad \text{i.e.,} \quad f_{T_2}(t_2) = \frac{1}{3}e^{-\frac{1}{3}t_2}, \quad t_2 > 0.$$

Since T_1, T_2 are independent, their joint pdf is given by

$$\begin{aligned} f_{T_1,T_2}(t_1,t_2) &= f_{T_1}(t_1) f_{T_2}(t_2) \\ &= \frac{1}{2}e^{-\frac{1}{2}t_1}\frac{1}{3}e^{-\frac{1}{3}t_2} \\ &= \frac{1}{6}e^{-\frac{1}{2}t_1}e^{-\frac{1}{3}t_2} \end{aligned}$$

for $t_1, t_2 > 0$. Let D be the support of the joint pdf $f_{T_1,T_2}(t_1,t_2)$, i.e., $D = \{(t_1,t_2)|0 < t_1, 0 < t_2\}$. Then it follows $P(T_2 < T_1) = P((T_1,T_2) \in A)$, where $A = \{(t_1,t_2) \in D|0 < t_1 < \infty, 0 < t_2 < t_1\}$. Hence

$$\begin{aligned} P(T_2 < T_1) &= P((T_1,T_2) \in A) \\ &= \int\int_A f_{T_1,T_2}(t_1,t_2)dt_1dt_2 \\ &= \int_0^\infty \left[\int_0^{t_1} \frac{1}{6}e^{-\frac{1}{2}t_1}e^{-\frac{1}{3}t_2}dt_2\right] dt_1 \\ &= \int_0^\infty \left[-\frac{1}{2}e^{-\frac{1}{2}t_1}e^{-\frac{1}{3}t_2}\right]_0^{t_1} dt_1 \\ &= \int_0^\infty \left(\frac{1}{2}e^{-\frac{1}{2}t_1} - \frac{1}{2}e^{-\frac{1}{2}t_1}e^{-\frac{1}{3}t_1}\right) dt_1 \\ &= 0.4 \end{aligned}$$

Question 3.1.5. ([4], No. 91) An insurance company insures a large number of drivers. Let X be the random variable representing the company's losses under collision insurance, and let Y represent the company's losses under liability insurance. X and Y have joint density function

$$f_{X,Y}(x,y) = \begin{cases} \frac{2x+2-y}{4}, & \text{for } 0 < x < 1,\ 0 < y < 2 \\ 0, & \text{otherwise,} \end{cases}$$

What is the probability that the total loss is at least 1?

SOLUTION. The question is to calculate $P(X+Y \geq 1)$. Let D be the support of the joint pdf $f_{X,Y}(x,y)$, i.e., $D = \{(x,y)|0 < x < 1, 0 < y < 2\}$. Then it follows $P(X+Y \geq 1) = P((X,Y) \in A)$, where $A = \{(x,y) \in D|x+y \geq 1\} = \{(x,y)|0 < x < 1, -x_1 \leq y < 2\}$. Hence

$$\begin{aligned} & P(X+Y \geq 1) \\ &= P((X,Y) \in A) \\ &= \int\int_A f_{X,Y}(x,y)dxdy \\ &= \int_0^1 \left[\int_{-x+1}^2 \frac{2x+2-y}{4}dy\right] dx \\ &= \int_0^1 \left[\frac{1}{2}xy + \frac{1}{2}y - \frac{1}{8}y^2\right]_{-x+1}^2 dx \\ &= \int_0^1 \left(\frac{5}{8}x^2 + \frac{3}{4}x + \frac{1}{8}\right) dx \\ &= \frac{17}{24} \end{aligned}$$

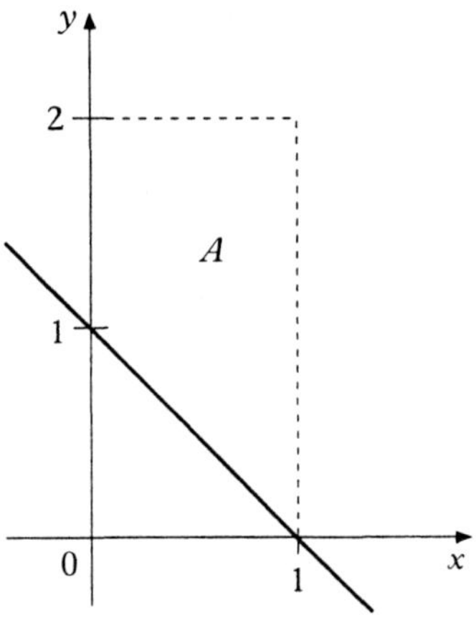

Question 3.1.6. ([4], No. 92) Two insurers provide bids on an insurance policy to a large company. The bids must be between 2,000 and 2,200. The company decides to accept the lower bid if the two bids differ by 20 or more. Otherwise, the company will consider the two bids further. Assume that the two bids are independent and are both uniformly distributed on the interval from 2,000 to 2,200.

Determine the probability that the company considers the two bids further.

SOLUTION. Let X and Y denote the two bids. The question is to determine $P(|X-Y|<20)$. We are given that X and Y are independent and $X, Y \sim U(2,000, 2,200)$, i.e.,

$$f_X(x) = f_Y(y) = \frac{1}{2,200 - 2,000} = \frac{1}{200}, \quad 2,000 < x, y < 2,200.$$

Thus the joint pdf of X and Y is given by

$$f_{X,Y}(x,y) = f_X(x)f_Y(y) = \begin{cases} \frac{1}{200^2}, & \text{for} \quad 2,000 < x, \ y < 2,200, \\ 0, & \text{otherwise.} \end{cases}$$

Let D be the support of the joint pdf $f_{X,Y}(x,y)$, i.e.,

$$D = \{(x,y)|2,000 < x < 2,200, \ 2,000 < y < 2,200\}.$$

Then it follows

$$P(|X-Y|<20) = P((X,Y) \in A),$$

where

$$\begin{aligned} A &= \{(x,y) \in D||x-y| < 20\} \\ &= \{(x,y)|2,020 < x < 2,200, x-20 < y < x+20\}. \end{aligned}$$

Since

$$\begin{aligned} A^c &= \{(x,y)|2,020 \leq x < 2,200, 2,000 < y \leq x-20\} \\ &\quad \cup \{(x,y)|2,000 < x \leq 2,180, x+20 \leq y < 2,200\} \\ &= A_1 \cup A_2, \end{aligned}$$

it holds

$$\begin{aligned}
&P(|X-Y|<20) \\
&= P((X,Y)\in A) \\
&= 1 - P((X,Y)\in A^c) \\
&= 1 - \{P((X,Y)\in A_1) + P((X,Y)\in A_2)\} \\
&= 1 - 2P((X,Y)\in A_1) \\
&= 1 - 2\int\int_{A_1} f_{X,Y}(x,y)dxdy \\
&= 1 - 2\int_{2,000}^{2,200}\left[\int_{x-20}^{2,000}\frac{1}{200^2}dy\right]dx \\
&= 1 - 2\int_{2,000}^{2,200}\left[\frac{y}{200^2}\right]_{x-20}^{2,000}dx \\
&= 1 - \frac{2}{200^2}\int_{2,000}^{2,200}(x-2,020)dx \\
&= 1 - \left(\frac{180}{200}\right)^2 = 0.19
\end{aligned}$$

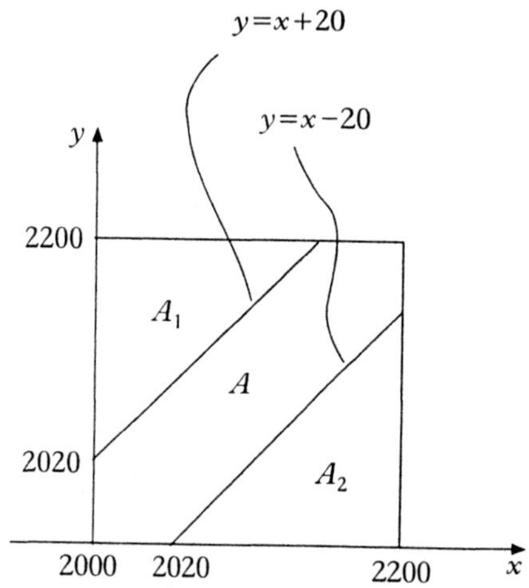

Question 3.1.7. ([4], No. 94) Let T_1 be the time between a car accident and reporting a claim to the insurance company. Let T_2 be the time between the report of the claim and payment of the claim. The joint density function of T_1 and T_2, $f_{T_1,T_2}(t_1,t_2)$, is constant over the region $0<t_1<6$, $0<t_2<6$, $t_1+t_2<10$, and zero other.

Determine $E[T_1+T_2]$, the expected time between a car accident and payment of the claim.

SOLUTION. Let

$$\begin{aligned}
D &= \text{support of} \quad f_{T_1,T_2}(t_1,t_2) \\
&= \{(t_1,t_2)|0<t_1<6, 0<t_2<6, t_1+t_2<10\}.
\end{aligned}$$

By the assumption of the question,

$$f_{T_1,T_2}(t_1,t_2) = k(= \text{a constant}) \quad \text{on} \quad D.$$

Thus

$$\begin{aligned}1 &= \int\int_D f_{T_1,T_2}(t_1,t_2)dt_1dt_2 = \int\int_D k dt_1 dt_2 \\ &= k(\text{area of} \quad D).\end{aligned}$$

Since

$$\begin{aligned}D &= \{(t_1,t_2)|0 < t_1 < 4, 0 < t_2 < 6\} \\ &\quad + \{(t_1,t_2)|4 \le t_1 < 6, 0 < t_2 < 10 - t_1\} \\ &= D_1 + D_2,\end{aligned}$$

the area of D is $6^2 - (1/2)(2)(2) = 34$.

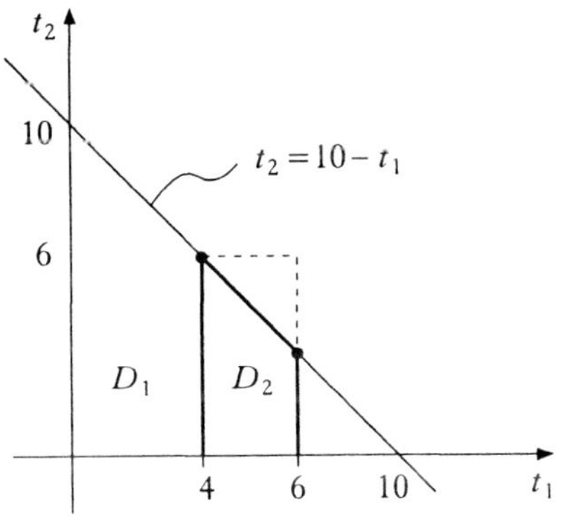

Thus we have $k = 1/34$ and so $f_{T_1,T_2}(t_1,t_2) = 1/34$. Then

$$\begin{aligned}E[T_1 + T_2] &= E[T_1] + E[T_2] = 2E[T_1] \\ &= 2\int\int_D t_1 f_{T_1,T_2}(t_1,t_2)dt_1dt_2 = 2\int\int_D \frac{t_1}{34}dt_1dt_2 \\ &= 2\left\{\int\int_{D_1}\frac{t_1}{34}dt_1dt_2 + \int\int_{D_2}\frac{t_1}{34}dt_1dt_2\right\} \\ &= 2\left\{\int_0^4\left[\int_0^6 \frac{t_1}{34}dt_2\right]dt_1 + \int_4^6\left[\int_0^{10-t}\frac{t_1}{34}dt_2\right]dt_1\right\} \\ &= 2\left\{\int_0^4 \frac{3t_1}{17}dt_1 + \int_4^6 \frac{1}{34}(10t_1 - t^2)dt_1\right\} \\ &= 5.7\end{aligned}$$

Question 3.1.8. ([4], No. 97) Let T_1 and T_2 represent the lifetimes in hours of two linked components in an electronic device. The joint density function for T_1 and T_2 is uniform over the region defined by $0 \le t_1 \le t_2 \le L$ where L is a positive constant.

Determine the expected value of the sum of the squares of T_1 and T_2.

SOLUTION. The question is to determine $E[T_1^2 + T_2^2]$. Let

$$D = \{(t_1, t_2) | 0 \le t_1 \le t_2 \le L\} = \{(t_1, t_2) | 0 \le t_1 \le L, t_1 \le t_2 \le L\}.$$

Since $(T_1, T_2) \sim U(D)$, the joint pdf is given by

$$f_{X,Y}(x, y) = \begin{cases} 1/(\text{area of } D), & \text{on } D, \\ 0, & \text{otherwise} \end{cases}$$

$$= \begin{cases} \frac{2}{L^2}, & \text{for } 0 \le t_1 \le L, t_1 \le t_2 \le L, \\ 0, & \text{otherwise.} \end{cases}$$

Thus

$$\begin{aligned} E[T_1^2 + T_2^2] &= \int\int_D (t_1^2 + t_2^2) f_{T_1,T_2}(t_1, t_2) dt_1 dt_2 \\ &= \int_0^L \left[\int_{t_1}^L (t_1^2 + t_2^2) \frac{2}{L^2} dt_2 \right] dt_1 \\ &= \int_0^L \frac{2}{L^2} \left[t_1^2 t_2 + \frac{1}{3} t_2^3 \right]_{t_1}^L dt_1 \\ &= \int_0^L \frac{2}{L^2} \left(t_1^2 L + \frac{L^3}{3} - \frac{4t_1^3}{3} \right) dt_1 \\ &= \frac{2}{L^2} \left[\frac{t_1^3 L}{3} + \frac{L^3 t_1}{3} - \frac{t_1^4}{3} \right]_0^L \\ &= \left(\frac{2}{L^2} \right) \left(\frac{L^4}{3} \right) = \frac{2}{3} L. \end{aligned}$$

Question 3.1.9. ([4], No. 101) The profit for a new product is given by $Z = 3X - Y - 5$. X and Y are independent random variable with $Var(X) = 1$ and $Var(Y) = 2$.

What is the variance of Z?

SOLUTION.

$$\begin{aligned}
Var(Z) &= Var(3X - Y - 5) \\
&= Var(3X - Y) \\
&= Var(3X) + Var(-Y) + 2Cov(3X, -Y) \\
&= 3^2 Var(X) + (-1)^2 Var(Y) + (2)(3)(-1)Cov(X, Y).
\end{aligned}$$

Since X and Y are independent, $Cov(X, Y) = 0$. Thus we have

$$Var(Z) = 3^2 Var(X) + (-1)^2 Var(Y) = 9 \times 1 + 2 \times 1 = 11.$$

Question 3.1.10. ([4], No. 104) A joint density function is given by

$$f_{X,Y}(x, y) = \begin{cases} kx, & \text{for } \ 0 < x < 1, 0 < y < 1, \\ 0, & \text{otherwise,} \end{cases}$$

where k is a constant.

What is $Cov(X, Y)$?

SOLUTION. To solve this question we use the formula

$$Cov(X, Y) = E[XY] - E[X]E[Y].$$

Let $D = \{(x, y) | 0 < x < 1, 0 < y < 1\}$. By the property of the joint pdf,

$$1 = \int\int_D f_{X,Y}(x, y)dxdy = \int_0^1 \int_0^1 kxdxdy = \frac{k}{2},$$

so that $k = 2$. Then

$$\begin{aligned}
E[X] &= \int\int_D x f_{X,Y}(x, y)dxdy \\
&= \int_0^1 \left[\int_0^1 (x)(2x)dy\right] dx \\
&= \int_0^1 \left[\int_0^1 2x^2 dy\right] dx \\
&= \frac{2}{3},
\end{aligned}$$

$$E[Y] = \int_0^1 \left[\int_0^1 (y)(2x)dy \right] dx$$
$$= \int_0^1 \left[\int_0^1 2xydy \right] dx = \frac{1}{2}$$

and

$$E[XY] = \int_0^1 \left[\int_0^1 (xy)(2x)dy \right] dx$$
$$= \int_0^1 \left[\int_0^1 2x^2ydy \right] dx = \frac{1}{3}.$$

Therefore

$$Cov(X,Y) = \frac{1}{3} - \left(\frac{2}{3}\right)\left(\frac{1}{2}\right) = 0.$$

Question 3.1.11. ([4], No. 105) Let X and Y be continuous random variable with joint density function

$$f_{X,Y}(x,y) = \begin{cases} \frac{8}{3}xy, & \text{for } 0 \le x \le 1, x \le y \le 2x, \\ 0, & \text{otherwise.} \end{cases}$$

Calculate $Cov(X,Y)$.

SOLUTION. Let $D = \{(x,y)|0 < x < 1, x < y < 2x\}$. Then

$$E[X] = \int\int_D xf_{X,Y}(x,y)dxdy$$
$$= \int_0^1 \left[\int_x^{2x} (x)\left(\frac{8}{3}xy\right) dy \right] dx = \frac{4}{5},$$
$$E[Y] = \int\int_D yf_{X,Y}(x,y)dxdy$$
$$= \int_0^1 \left[\int_x^{2x} (y)\left(\frac{8}{3}xy\right) dy \right] dx = \frac{56}{45}$$

and

$$E[XY] = \int\int_D xyf_{X,Y}(x,y)dxdy$$
$$= \int_0^1 \left[\int_x^{2x} (xy)\left(\frac{8}{3}xy\right) dy \right] dx = \frac{28}{27}.$$

Therefore

$$Cov(X, Y) = \frac{28}{27} - \left(\frac{4}{5}\right)\left(\frac{56}{45}\right) = 0.04.$$

Question 3.1.12. ([4], No. 107) Let X denote the size of a surgical claim and let Y denote the size of the associated hospital claim. An actuary is using a model in which $E[X] = 5$, $E[X^2] = 27.4$, $E[Y] = 7$, $E[Y^2] = 51.4$, and $Var(X + Y) = 8$.

Let $C_1 = X+Y$ denote the size of the combined claims before the application of a 20% surcharge on the hospital portion of the claim, and C_2 denote the the size of the combined claims after the application of that surcharge.

Calculate $Cov(C_1, C_2)$.

SOLUTION. Since $C_2 = X + 1.2Y$, from the definition of the covariance, we see that

$$\begin{aligned}
&Cov(C_1, C_2) \\
&= Cov(X + Y, X + 1.2Y) \\
&= Cov(X, X) + Cov(X, 1.2Y) + Cov(Y, X) + Cov(Y, 1.2Y) \\
&= Var(X) + 1.2Cov(X, Y) + Cov(Y, X) + 1.2Var(Y) \\
&= Var(X) + 1.2Var(Y) + 2.2Cov(X, Y).
\end{aligned}$$

Here

$$Var(X) = E[X^2] - E[X]^2 = 27.4 - 5^2 = 2.4$$

and

$$Var(Y) = E[Y^2] - E[Y]^2 = 51.4 - 7^2 = 2.4.$$

Thus

$$\begin{aligned}
Var(X + Y) &= Var(X) + Var(Y) + 2Cov(X, Y) \\
&= 2.4 + 2.4 + 2Cov(X, Y) \\
&= 4.8 + 2Cov(X, Y).
\end{aligned}$$

But since $Var(X+Y) = 8$ by the assumption,

$$8 = 4.8 + 2Cov(X, Y)$$

so that $Cov(X, Y) = 1.6$. Therefore,

$$\begin{aligned} Cov(C_1, C_2) &= Var(X) + 1.2Var(Y) + 2.2Cov(X, Y) \\ &= 2.4 + (1.2)(2.4) + (2.2)(1.6) = 8.8. \end{aligned}$$

Question 3.1.13. ([4], No. 108) A device containing two key components fails when, and only when, both components fail. The lifetimes, T_1 and T_2, of these components are independent with common density function $f(t) = e^t$, $t > 0$. The cost, X, of operating the device until failure is $2T_1 + T_2$.

Find the density function of X for $x > 0$.

SOLUTION. Since T_1 and T_2 are independent, they have the joint pdf

$$f_{T_1,T_2}(t_1, t_2) = f_{T_1}(t_1) f_{T_2}(t_2) = \begin{cases} e^{t_1} e^{t_2}, & \text{for} \quad 0 < t_1, 0 < t_2, \\ 0, & \text{otherwise.} \end{cases}$$

Let $D = \{(t_1, t_2) | 0 < t_1, 0 < t_2\}$. Then

$$\begin{aligned} F_X(x) &= P(X \le x) = P(2T_1 + T_2 \le x) \\ &= P((T_1, T_2) \in A), \end{aligned}$$

where

$$\begin{aligned} A &= \{(t_1, t_2) \in D | 2t_1 + t_2 \le x\} \\ &= \{(t_1, t_2) | 0 < t_1 \le 1/2(x - t_2), 0 < t_2 \le x\}. \end{aligned}$$

Thus

$$\begin{aligned} F_X(x) &= \int\int_A f_{T_1,T_2}(t_1, t_2) dt_1 dt_2 \\ &= \int_0^x \left[\int_0^{1/2(x-t_1} e^{t_1} e^{t_2} dt_1 \right] dt_2 \\ &= 1 - 2e^{-x/2} + e^{-x} \end{aligned}$$

for $x > 0$. Therefore we have

$$f_X(x) = F'_X(x) = \left[1 - 2e^{-x/2} + e^{-x}\right]' = e^{-x/2} + e^{-x}.$$

Question 3.1.14. ([4], No. 109) A company offers earthquake insurance. Annual premiums are modeled by an exponential random variable with mean 2. Annual claims are modeled by an exponential random variable with mean 1. Premiums and claims are independent. Let X denote the ratio of claims to premiums.

What is the density function of X?

Solution. Let U and V denote annual claims and annual premiums, respectively. We are given that $X = U/V$ and

$$\begin{aligned} U &\sim E(1), \quad \text{i.e.,} \quad f_U(u) = e^{-u}, \quad u > 0, \\ V &\sim E(1/2), \quad \text{i.e.,} \quad f_V(v) = \frac{1}{2}e^{-\frac{1}{2}v}, \quad v > 0. \end{aligned}$$

Since U and V are independent, they has the following joint pdf

$$f_{U,V}(u,v) = f_U(u)f_V(v) = \frac{1}{2}e^{-u}e^{-\frac{1}{2}v}.$$

for $u, v > 0$. Let $D = \{(u,v)|0 < u, 0 < v\}$. Then

$$\begin{aligned} F_X(x) = P(X \le x) &= P\left(\frac{U}{V} \le x\right) \\ &= P((U,V) \in A), \end{aligned}$$

where

$$\begin{aligned} A &= \{(u,v) \in D | \frac{u}{v} \le x\} \\ &= \{(u,v)|0 < u \le xv, 0 < v < \infty\}. \end{aligned}$$

Thus

$$\begin{aligned}F_X(x) &= \int\int_A f_{U,V}(u,v)dudv\\ &= \int_0^\infty \left[\int_0^{xv} \frac{1}{2}e^{-u}e^{-\frac{1}{2}v}du\right]dv\\ &= \int_0^\infty \left(-\frac{1}{2}e^{-v(x+\frac{1}{2})} + \frac{1}{2}e^{-\frac{v}{2}}\right)dv\\ &= -\frac{1}{2x+1} + 1.\end{aligned}$$

Therefore we have

$$f_X(x) = F'_X(x) = \left[-\frac{1}{2x+1} + 1\right]' = \frac{2}{(2x+1)^2}.$$

Question 3.1.15. ([4], No. 117) A company is reviewing tornado damage claims under a farm insurance policy. Let X be the portion of a claim representing damage to the house and Y be the portion of the same claim representing damage to the rest of the property. The joint density function X and Y is

$$f_{X,Y}(x,y) = \begin{cases} 6[1-(x+y)], & \text{for} \quad x > 0, y > 0, x+y < 1 \\ 0, & \text{otherwise.} \end{cases}$$

Determine the probability that the portion of a claim representing damage to the house is less than 0.2.

SOLUTION. The question is to determine $P(X < 0.2)$. Let D be the support of the joint pdf $f_{X,Y}(x,y)$, i.e., $D = \{(x,y)|x > 0, y > 0, x+y < 1\} = \{(x,y)|0 < x, 0 < y < -x+1\}$. Then it follows $P(X < 0.2) = P((X,Y) \in A)$, where $A = \{(x,y) \in D|x < 0.2\} = \{(x,y)|0 < x < 0.2, 0 <$

$y < -x + 1\}$. Hence

$$\begin{aligned}
P(X < 0.2) &= \int\int_A f_{X,Y}(x,y)dxdy \\
&= \int_0^{0.2} \left[\int_0^{-x+1} 6[1-(x+y)] \right] dx \\
&= 6 \int_0^{0.2} \left[y - xy - \frac{1}{2}y^2 \right]_0^{-x+1} dx \\
&= 6 \int_0^{0.2} \frac{1}{2}(1-x)^2 dx \\
&= [-(1-x)^3]_0^{0.2} = -(0.8)^3 + 1 = 0.488.
\end{aligned}$$

Question 3.1.16. ([4], No. 118) Let X and Y be continuous random variables with joint density function

$$f_{X,Y}(x,y) = \begin{cases} 15y, & \text{for } \quad x^2 \le y \le x \\ 0, & \text{otherwise.} \end{cases}$$

Find the marginal density function $f_Y(y)$ of Y.

Solution. Let D be the support of the joint pdf $f_{X,Y}(x,y)$.Then

$$\begin{aligned}
D &= \{(x,y)|x^2 \le y \le x\} \\
&= \{(x,y)|0 \le x \le 1, x^2 \le y \le x\} \\
&= \{(x,y)|0 \le y \le 1, y \le x \le \sqrt{y}\}.
\end{aligned}$$

By the definition of the marginal density function, we have

$$\begin{aligned}
f_Y(y) &= \int_{-\infty}^{\infty} f_{X,Y}(x,y)dx \\
&= \int_y^{\sqrt{y}} 15ydx \\
&= [15yx]_{x=y}^{x=\sqrt{y}} \\
&= 15y(\sqrt{y} - y), \quad 0 \le y \le 1.
\end{aligned}$$

Question 3.1.17. ([4], No. 120) An insurance policy is written to cover a loss X where X has density function

$$f_X(x) = \begin{cases} \frac{3}{8}x^2, & \text{for} \quad 0 \le x \le 2 \\ 0, & \text{otherwise.} \end{cases}$$

The time (in hours) to process a claim of size x, where $0 \le x \le 2$, is uniformly distributed on the interval from x to $2x$.

Calculate the probability that a randomly chosen claim on this policy is processed in three hours or more.

SOLUTION. Let T be the time to process a claim. The question is to calculate $P(T \ge 3)$. We are given that T is independent of X and $T \sim U(x, 2x)$, i.e.,

$$f_T(t) = \begin{cases} \frac{1}{2x-x} = \frac{1}{x}, & \text{for} \quad x \le t \le 2x \\ 0, & \text{otherwise.} \end{cases}$$

Since X and T are independent, the joint pdf is given by

$$f_{X,T}(x,t) = f_X(x)f_T(t) = \frac{3}{8}x^2\frac{1}{x} = \frac{3}{8}x$$

for $0 \le x \le 2, x \le t \le 2x$. Let D be the support of $f_{X,T}(x,t)$, i.e., $D = \{(x,t)|0 \le x \le 2, x \le t \le 2x\}$. Then $P(T \ge 3) = P((X,T) \in A)$, where

$$\begin{aligned} A &= \{(x,t) \in D | t \ge 3\} \\ &= \{(x,t)|3 \le t \le 4, t/2 < x \le 2\}. \end{aligned}$$

Thus

$$\begin{aligned} P(T \ge 3) &= \int\int_A f_{X,T}(x,t)dxdt \\ &= \int_3^4 \left[\int_{t/2}^2 \frac{3}{8}xdx\right] dt \\ &= \int_3^4 \left(\frac{12}{16} - \frac{3}{64}t^2\right) dt \\ &= \frac{11}{64}. \end{aligned}$$

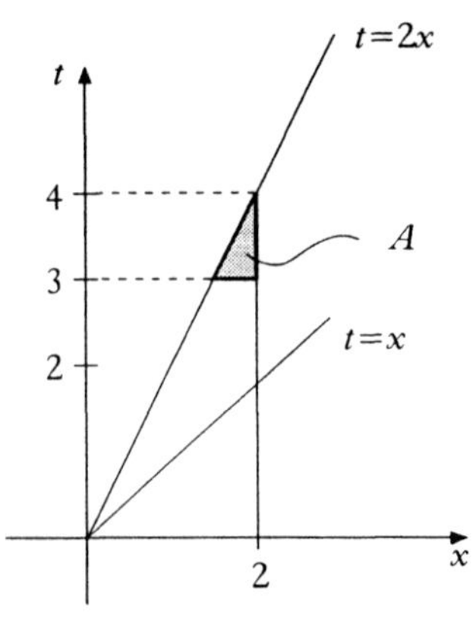

Question 3.1.18. ([4], No. 121) Let X represent the age of an insured automobile involved in an accident. Let Y represent the length of time the owner has insured the automobile at the time of the accident.
X and Y have joint probability density function

$$f_{X,Y}(x,y) = \begin{cases} \frac{1}{64}(10 - xy^2), & \text{for} \quad 2 \le x \le 10, 0 \le y \le 1 \\ 0, & \text{otherwise.} \end{cases}$$

Calculate the expected age of an insured automobile involved in an accident.

SOLUTION. We want to calculate $E[X]$. To to so we need to determine the marginal pdf $f_X(x)$ of X.

$$\begin{aligned} f_X(x) &= \int_{-\infty}^{\infty} f_{X,Y}(x,y)dy \\ &= \int_0^1 \frac{1}{64}(10 - xy^2)dy \\ &= \frac{1}{64}\left[10y - \frac{xy^3}{3}\right]_0^1 \\ &= \frac{1}{64}\left(10 - \frac{x}{3}\right). \end{aligned}$$

Therefore

$$\begin{aligned} E[X] &= \int_{-\infty}^{\infty} x f_X(x)dx \\ &= \int_2^{10} x\frac{1}{64}\left(10 - \frac{x}{3}\right)dx \\ &= \frac{1}{64}\int_2^{10}\left(10x - \frac{x^2}{3}\right)dx \\ &= \frac{1}{64}\left[5x^2 - \frac{x^3}{9}\right]_2^{10} \\ &= 5.778. \end{aligned}$$

Alternatively, this question can be solved by the method used to solve Question 3.1.10 or 3.1.11 in where it is not necessary to determine the marginal pdf $f_X(x)$ of X. That is, $E[X]$ is given directly by the following formula.

$$E[X] = \int\int_D x f_{X,Y}(x,y)dxdy,$$

where D is the support of $f_{X,Y}(x,y)$. In fact,

$$\begin{aligned}
E[X] &= \int\int_D x f_{X,Y}(x,y)dxdy \\
&= \int_2^{10}\left[\int_0^1 x\frac{1}{64}(10-xy^2)dy\right]dx \\
&= \frac{1}{64}\int_2^{10}\left[\int_0^1 (10x-x^2y^2)dy\right]dx \\
&= \frac{1}{64}\int_2^{10}\left[10xy-\frac{x^2y^3}{3}\right]_{y=0}^{y=1}dx \\
&= \frac{1}{64}\int_2^{10}\left(10x-\frac{x^2}{3}\right)dx \\
&= \frac{1}{64}\left[5x^2-\frac{x^3}{9}\right]_2^{10} \\
&= 5.778.
\end{aligned}$$

Question 3.1.19. ([4], No. 122) A device contains two circuits. The second circuit is a backup for the first, so the second is used only when the first has failed. The device fails when and only when the second circuit fails. Let X and Y be the times at which the first and second circuits fail, respectively. X and Y have joint probability density function

$$f_{X,Y}(x,y) = \begin{cases} 6e^{-x}e^{-2y}, & \text{for } 0 < x < y < \infty \\ 0, & \text{otherwise.} \end{cases}$$

What is the expected time at which the device fails?

SOLUTION. The question is to find $E[Y]$. To to so we need to determine the marginal pdf $f_Y(y)$ of Y.

$$\begin{aligned} f_Y(y) &= \int_{-\infty}^{\infty} f_{X,Y}(x,y)dx \\ &= \int_0^y 6e^{-x}e^{-2y}dx \\ &= 6e^{-2y}\int_0^y e^{-x}dx \\ &= 6e^{-2y}\left[-e^{-x}\right]_0^y \\ &= 6e^{-2y} - 6e^{-3y},\ 0 < y < \infty. \end{aligned}$$

Therefore

$$\begin{aligned} E[Y] &= \int_{-\infty}^{\infty} y f_Y(y)dy \\ &= \int_0^{\infty} y\left(6e^{-2y} - 6e^{-3y}\right)dy \\ &= \frac{6}{2}\int_0^{\infty} 2ye^{-2y}dy - 2\int_0^{\infty} 3ye^{-3y}dy \\ &= \left(\frac{6}{2}\right)\left(\frac{1}{2}\right) - (2)\left(\frac{1}{3}\right) = \frac{5}{6}. \end{aligned}$$

Here we used the result that if $T \sim E(\lambda)$, then

$$\frac{1}{\lambda} = E[T] = \int_0^{\infty} \lambda t e^{-\lambda t}dt.$$

Alternatively, this question also can be solved by the following formula.

$$E[X] = \int\int_D x f_{X,Y}(x,y)dxdy,$$

where D is the support of $f_{X,Y}(x,y)$.

3.2. Conditional distributions

Question 3.2.1. ([4], No. 95) Let X and Y be independent random variables with common moment generating function $M(t) = e^{t^2/2}$.
Let $W = X + Y$ and $Z = Y - X$.

Determine the joint moment generating function, $M_{W,Z}(t_1, t_2)$, of W and Z.

SOLUTION. By the assumption of the question, it follows that

$$\begin{aligned} M_{W,Z}(t_1,t_2) &= E\left[e^{t_1 W + t_2 Z}\right] \\ &= E\left[e^{t_1(X+Y)+t_2(Y-X)}\right] \\ &= E\left[e^{(t_1-t_2)X}e^{(t_1+t_2)Y}\right] \\ &= E\left[e^{(t_1-t_2)X}\right]E\left[e^{(t_1+t_2)Y}\right] \\ &= M_X(t_1-t_2)M_Y(t_1+t_2) \\ &= e^{\frac{1}{2}(t_1-t_2)^2}e^{\frac{1}{2}(t_1+t_2)^2} \\ &= e^{t_i^2+t_2^2}. \end{aligned}$$

Question 3.2.2. ([4], No. 106) Let X and Y denote the values of two stocks at the end of a five-year period. X is uniformly distributed on the interval (0,12). Given $X = x$, Y is uniformly distributed on the interval $(0, x)$.

Determine $Cov(X, Y)$ according to this model.

SOLUTION. To use the formula

$$Cov(X,Y) = E[XY] - E[X]E[Y],$$

we need to determine the joint pdf $f_{X,Y}(x,y)$ of X and Y. We are given that

$$X \sim U(0,12), \quad \text{i.e.,} \quad f_X(x) = \frac{1}{12}, \quad 0 < x < 12,$$

$$Y|_{X=x} \sim U(0,x), \quad \text{i.e.,} \quad f_{Y|X}(y|X=x) = \frac{1}{x}, \quad 0 < y < x.$$

By Theorem 3.2.3, it holds

$$f_{X,Y}(x,y) = f_{Y|X}(y|X=x)f_X(x) = \left(\frac{1}{x}\right)\left(\frac{1}{12}\right) = \frac{1}{12x}$$

for $0 < x < 12,\ 0 < y < x$, i.e.,

$$f_{X,Y}(x,y) = \begin{cases} \frac{1}{12x}, & \text{for} \quad 0 < x < 12,\ 0 < y < x \\ 0, & \text{otherwise.} \end{cases}$$

Let $D = \{(x,y)|0 < x < 12,\ 0 < y < x\}$. Then we have

$$E[X] = \int\int_D x f_{X,Y}(x,y)dxdy = \int_0^{12}\left[\int_0^x x\frac{1}{12x}dx\right]dy = 6,$$

$$E[Y] = \int\int_D y f_{X,Y}(x,y)dxdy = \int_0^{12}\left[\int_0^x y\frac{1}{12x}dx\right]dy = 3$$

and

$$E[XY] = \int\int_D xy f_{X,Y}(x,y)dxdy = \int_0^{12}\left[\int_0^x xy\frac{1}{12x}dx\right]dy = 24.$$

Therefore, it follows

$$Cov(X,Y) = E[XY] - E[X]E[Y] = 24 - (6)(3) = 6.$$

Question 3.2.3. ([4], No. 111) Once a fire is reported to a fire insurance company, the company makes an initial estimate, X, of the amount it will pay to the claimant for the fire loss. When the claim in finally settled, the company pays an amount, Y, to the claimant. The company has determined that X and Y have the joint density function

$$f_{X,Y}(x,y) = \frac{2}{x^2(x-1)}y^{-(2x-1)/(x-1)}, \quad x, y > 1.$$

Given that the initial claim estimated by the company is 2, determine the probability that the final settlement amount is between 1 and 3.

SOLUTION. The question is to determine $P(1 < Y < 3|X = 2)$. To do this we use again Theorem 3.2.3.

$$\begin{aligned} P(1 < Y < 3|X = 2) &= \int_1^3 f_{Y|X}(y|X = 2)dy \\ &= \int_1^3 \frac{f_{X,Y}(2, y)}{f_X(2)} dy. \end{aligned}$$

From the assumption, it follows that for $1 < y < \infty$,

$$f_{X,Y}(2, y) = \frac{2}{2^2(2-1)} y^{-(2\cdot 2-1)/(2-1)} = \frac{1}{2} y^{-3}$$

and hence

$$f_X(2) = \int_{-\infty}^{\infty} f_{X,Y}(2, y)dy = \int_1^{\infty} \frac{1}{2} y^{-3} dy = \frac{1}{4}.$$

Therefore we have

$$\begin{aligned} P(1 < Y < 3|X = 2) &= \int_1^3 \frac{\frac{1}{2}y^{-3}}{\frac{1}{4}} dy = \int_1^3 2y^{-3}dy \\ &= \left[-\frac{1}{y^2} \right]_1^3 = 1 - \frac{1}{9} = \frac{8}{9}. \end{aligned}$$

Question 3.2.4. ([4], No. 112) A company offers a basic life insurance policy to its employees, as well as a supplemental life insurance policy. To purchase the supplemental policy, an employee must first purchase the basic policy.

Let X denote the proportion of employees who purchase the basic policy, and Y denote the proportion of employees who purchase the supplemental policy. Let X and Y have the joint density function $f_{X,Y}(x, y) = 2(x + y)$ on the region where the density is positive.

Given that 10% of the employees buy the basic policy, what is the probability that fewer than 5% buy the supplemental policy?

SOLUTION. The question is to determine $P(Y < 0.05|X = 0.1)$. To do this we use Theorem 3.2.3 again.

$$\begin{aligned}P(Y < 0.05|X = 0.1) &= \int_0^{0.05} f_{Y|X}(y|X = 0.1)dy\\ &= \int_0^{0.05} \frac{f_{X,Y}(0.1, y)}{f_X(0.1)}dy.\end{aligned}$$

From the assumption, we are given that

$$f_{X,Y}(x, y) = \begin{cases} 2(x + y), & \text{for } 0 < x < 1,\ 0 < y < x \\ 0, & \text{otherwise.} \end{cases}$$

Thus

$$f_{X,Y}(0.1, y) = 2(0.1 + y) = 0.2 + 2y$$

for $0 < y < x$. And since we have

$$f_X(x) = \int_{-\infty}^{\infty} f_{X,Y}(x, y)dy = \int_0^x 2(x + y)dy = 3x^2,$$

we see that

$$f_X(0.1) = 3(0.1)^2 = 0.03.$$

Therefore

$$\begin{aligned}P(Y < 0.05|X = 0.1) &= \int_0^{0.05} \frac{f_{X,Y}(0.1, y)}{f_X(0.1)}dy\\ &= \int_0^{0.05} \frac{0.2 + 2y}{0.03}dy\\ &= \int_0^{0.05} \left(\frac{20}{3} + \frac{200}{3}y\right)dy\\ &= \frac{5}{12} = 0.417.\end{aligned}$$

Question 3.2.5. ([4], No. 114) A diagnostic test for the presence of a disease has two possible outcomes: 1 for disease present and 0 for disease not present. Let X denote the disease state of a patient, and let Y denote the outcome of the diagnostic test. The joint probability function of X and

Y is given by:

$$P(X=0,Y=0)=0.800$$

$$P(X=1,Y=0)=0.050$$

$$P(X=0,Y=1)=0.025$$

$$P(X=1,Y=1)=0.125$$

Calculate $Var(Y|X=1)$.

SOLUTION. We use the following formula

$$Var(Y|X=1)=E[Y^2|X=1]-E[Y|X=1]^2.$$

Here

$$\begin{aligned}E[Y^2|X=1]&=(0)^2P(Y=0|X=1)+(1)^2P(Y=1|X=1)\\&=P(Y=1|X=1)\\&=\frac{P(X=1,Y=1)}{P(X=1)}\\&=\frac{P(X=1,Y=1)}{P(X=1,Y=0)+P(X=1,Y=1)}\\&=\frac{0.125}{0.050+0.125}=0.714\end{aligned}$$

and

$$\begin{aligned}E[Y|X=1]&=(0)P(Y=0|X=1)+(1)P(Y=1|X=1)\\&=P(Y=1|X=1)\\&=0.714.\end{aligned}$$

Therefore we have

$$\begin{aligned}Var(Y|X=1)&=E[Y^2|X=1]-E[Y|X=1]^2\\&=0.714-0.714^2=0.20.\end{aligned}$$

Question 3.2.6. ([4], No. 115) The stock prices of two companies at the end of any given year are modeled with random variables X and Y that follow a distribution with joint density function

$$f_{X,Y}(x,y) = \begin{cases} 2x, & \text{for} \quad 0 < x < 1,\ x < y < x+1 \\ 0, & \text{otherwise.} \end{cases}$$

What is the conditional variance of Y given that $X = x$?

SOLUTION. To find $Var(Y|X = x)$, we use the formula

$$Var(Y|X = x) = E[Y^2|X = x] - E[Y|X = x]^2.$$

The first term of the right hand side is given by

$$\begin{aligned} E[Y^2|X = x] &= \int_{-\infty}^{\infty} y^2 f_{Y|X}(y|X = x)dy \\ &= \int_{x}^{x+1} y^2 \frac{f_{X,Y}(x,y)}{f_X(x)} dy \end{aligned}$$

Since

$$\begin{aligned} f_X(x) &= \int_{-\infty}^{\infty} f_{X,Y}(x,y)dy \\ &= \int_{x}^{x+1} 2x dy = 2x, \end{aligned}$$

we have

$$\begin{aligned} E[Y^2|X = x] &= \int_{x}^{x+1} y^2 \frac{2x}{2x} dy \\ &= \int_{x}^{x+1} y^2 dy \\ &= x^2 + x + \frac{1}{3}. \end{aligned}$$

Similarly, the second term is given by

$$\begin{aligned} E[Y|X = x] &= \int_{x}^{x+1} y \frac{2x}{2x} dy \\ &= \int_{x}^{x+1} y dy \\ &= x + \frac{1}{2}. \end{aligned}$$

Therefore

$$\begin{aligned} Var(Y|X = x) &= E[Y^2|X = x] - E[Y|X = x]^2 \\ &= x^2 + x + \frac{1}{3} - \left(x + \frac{1}{2}\right)^2 \\ &= \frac{1}{12}. \end{aligned}$$

Question 3.2.7. ([4], No. 116) An actuary determines that the annual numbers of tornados in counties P and Q are jointly distributed as follows:

		Annual number of tornadoes in county Q			
		0	1	2	3
Annual number of tornadoes in county P	0	0.12	0.06	0.05	0.02
	1	0.13	0.15	0.12	0.03
	2	0.05	0.15	0.10	0.02

Calculate the conditional variance of the annual number of tornadoes in county Q, given that there no tornadoes in county P.

SOLUTION. Let N_P and N_Q be the annual numbers of tornadoes in counties P and Q, respectively. The question is to find $Var(N_Q|N_P = 0)$. We use the formula

$$Var(N_Q|N_P = 0) = E[N_Q^2|N_P = 0] - E[N_Q|N_P = 0]^2$$

Here

$$\begin{aligned}E[N_Q^2|N_P=0] &= \sum_{k=0}^{3} k^2 P_{N_Q|N_P}(k|N_P=0)\\ &= \sum_{k=0}^{3} k^2 \frac{P(N_Q=k, N_P=0)}{P(N_P=0)}\\ &= \frac{\sum_{k=0}^{3} k^2 P(N_Q=k, N_P=0)}{P(N_P=0)}\\ &= \frac{(0)^2(0.12)+(1)^2(0.06)+(2)^2(0.05)+(3)^2(0.02)}{0.12+0.06+0.05+0.02}\\ &= 1.76.\end{aligned}$$

Similarly we get

$$\begin{aligned}E[N_Q|N_P=0] &= \sum_{k=0}^{3} k P_{N_Q|N_P}(k|N_P=0)\\ &= \frac{(0)(0.12)+(1)(0.06)+(2)(0.05)+(3)(0.02)}{0.12+0.06+0.05+0.02}\\ &= 0.88.\end{aligned}$$

Therefore we have

$$\begin{aligned}Var(N_Q|N_P=0) &= E[N_Q^2|N_P=0] - E[N_Q|N_P=0]^2\\ &= 1.76 - 0.88^2 = 0.99.\end{aligned}$$

Question 3.2.8. ([**4**], No. 119) An auto insurance policy will pay for damage to both the policyholder's car and the other driver's car in the event that the policyholder is responsible for an accident. The size of the payment for damage to the policyholder's car, X, has a marginal density function of 1 for $0 < x < 1$. Given $X = x$, the size of the payment for damage to the other driver's car, Y, has conditional density 1 for $x < y < x + 1$.

If the policyholder is responsible for an accident, what is the probability that the payment for damage to the other driver's car will be greater than 0.5?

SOLUTION. The question is to find $P(Y > 0.5)$. We are given that

$$f_X(x) = 1 \quad \text{for} \quad 0 < x < 1$$

and

$$f_{Y|X}(y|X = x) = 1 \quad \text{for} \quad x < y < x + 1.$$

From these, we get

$$f_{X,Y}(x, y) = f_{Y|X}(y|X = x)f_X(x) = 1 \times 1 = 1$$

for $0 < x < 1,\ x < y < x + 1$. Let $D = \{(x, y)|0 < x < 1,\ x < y < x + 1\}$. Then

$$\begin{aligned} P(Y > 0.5) &= 1 - P(Y \le 0.5) \\ &= 1 - P((X, Y) \in A), \end{aligned}$$

where

$$\begin{aligned} A &= \{(x, y) \in D|y \le 0.5\} \\ &= \{(x, y)|0 < x < 1, x < y < x + 1, y > 0.5\} \\ &= \{(x, y)|0 < x < \frac{1}{2}, x < y < \frac{1}{2}\}. \end{aligned}$$

Thus

$$\begin{aligned} P(Y > 0.5) &= 1 - \int\int_A f_{X,Y}(x, y)dxdy \\ &= 1 - \int_0^{\frac{1}{2}} \left[\int_x^{\frac{1}{2}} 1dy\right] dx \\ &= 1 - \int_0^{\frac{1}{2}} \left(\frac{1}{2} - x\right) dx \\ &= 1 - \frac{1}{8} = \frac{7}{8}. \end{aligned}$$

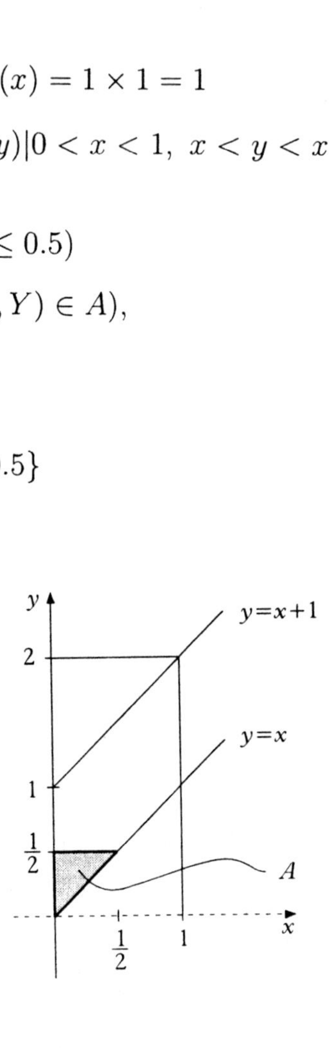

CHAPTER 4

Functions of Random Variables

4.1. Distributions of functions of random variables

Question 4.1.1. ([4], No. 37) The lifetime of a printer costing 200 is exponentially distributed with mean 2 years. The manufacturer agrees to pay a full refund to a buyer if the printer fails during the first year following its purchase, and a one-half refund if it fails during the second year.

If the manufacturer sells 100 printers, how much should it expect to pay in refunds?

SOLUTION. Let T denote the lifetime of a printer, and let Y_i be the refund for i-th printer, $i = 1, 2, \cdots, 100$. The question is to find $E[\sum_{i=1}^{100} Y_i]$. We are given that

$$Y_i = \begin{cases} 200, & \text{w.p. } \ P(T \le 1) \\ 100, & \text{w.p. } \ P(1 \le T \le 2) \\ 0, & \text{w.p. } \ P(T > 3) \end{cases}$$

where $T \sim E(1/2)$, i.e., $f_T(t) = 1/2e^{-t/2}$ for $t \geq 0$. Thus we have

$$\begin{aligned}
E[Y_i] &= (200)P(T \leq 1) + (100)P(1 \leq T \leq 2) + (0)P(T > 3) \\
&= 200 \int_0^1 \frac{1}{2} e^{-\frac{t}{2}} dt + 100 \int_1^2 \frac{1}{2} e^{-\frac{t}{2}} dt \\
&= 200 \left[-e^{-\frac{t}{2}} \right]_0^1 + 100 \left[-e^{-\frac{t}{2}} \right]_1^2 \\
&= 102.56.
\end{aligned}$$

so that

$$E\left[\sum_{i=1}^{100} Y_i\right] = \sum_{i=1}^{100} E\left[Y_i\right] = 100 \times 102.56 = 10,256.$$

Question 4.1.2. ([4], No. 99) An insurance policy pays a total medical benefit consisting of two parts for each claim. Let X represent the part of the benefit that is paid to the surgeon, and let Y represent the part that is paid to the hospital. The variance of X is 5,000, the variance of Y is 10,000, and the variance of the total benefit, $X + Y$, is 17,000.
Due to increasing medical costs, the company that issues the policy decides to increase X by a flat amount of 100 per claim and to increase Y by 10% per claim.

Calculate the variance of the total benefit after these revisions have been made.

Solution. The question is to calculate $Var(X + 100 + 1.1Y)$. We are given that

$$\begin{aligned}
Var(X) &= 5,000 \\
Var(Y) &= 10,000 \\
Var(X + Y) &= 17,000.
\end{aligned}$$

From these we have

$$\begin{aligned} 17,000 &= Var(X+Y) \\ &= Var(X) + Var(Y) + 2Cov(X,Y) \\ &= 5,000 + 10,000 + 2Cov(X,Y) \end{aligned}$$

so that $Cov(X,Y) = 1,000$. Therefore we get

$$\begin{aligned} &Var(X + 100 + 1.1Y) \\ &= Var(X + 1.1Y) \\ &= Var(X) + Var(1.1Y) + 2Cov(X, 1.1Y) \\ &= Var(X) + (1.1)^2 Vay(Y) + (2)(1.1)Cov(X,Y) \\ &- 5,000 + (1.1)^2(10,000) + (2)(1.1)(1,000) \\ &= 19,300. \end{aligned}$$

Question 4.1.3. ([4], No. 102) A company has two electric generators. The time until failure for each generator follows an exponential distribution with mean 10. The company will begin using the second generator immediately after one fails.

What is the variance of the total time that the generators produce electricity?

SOLUTION. Let X and Y denote the times that two generators produce electricity. The question is to find $Var(X+Y)$. Since $X, Y \sim E(\frac{1}{10})$, $Var(X) = Var(Y) = 100$. Because X and Y are independent, it follows

$$\begin{aligned} Var(X+Y) &= Var(X) + Var(Y) \\ &= 100 + 100 = 200. \end{aligned}$$

4.2. The central limit theorem

Question 4.2.1. ([4], No. 81) Claims filed under auto insurance policies follow a normal distribution with mean 19,400 and standard deviation 5,000.

What is the probability that the average of 25 randomly selected claims exceeds 20,000?

SOLUTION. Let $X_1, X_2, \cdots, X_{25}$ be 25 randomly selected claims and, let

$$\overline{X} = \frac{1}{25}\sum_{k=1}^{25} X_k.$$

The question is to find $P(\overline{X} > 20,000)$. Since, for all $k = 1, 2, \cdots, 25$, $X_k \sim N(19,400, 5,000^2)$ and they are independent, we have

$$E[\overline{X}] = \frac{1}{25}\sum_{k=1}^{25} E[X_k] = \frac{1}{25} \times 25 \times 19,400 = 19,400.$$

and

$$Var(\overline{X}) = \frac{1}{25^2}\sum_{k=1}^{25} Var(X_k) = \frac{1}{25^2} \times 25 \times 5,000^2 = 1,000^2.$$

By the central limit theorem, we see that

$$Z \equiv \frac{\overline{X} - E[\overline{X}]}{\sqrt{Var(\overline{X})}} \sim N(0, 1^2) \quad \text{approximately}$$

Thus, by the standard normal table, we have

$$\begin{aligned}
P(\overline{X} > 20,000) &= P\left(\frac{\overline{X} - 19,400}{1,000} > \frac{20,000 - 19,400}{1,000}\right) \\
&= P(Z > 0.6) \\
&= 1 - P(Z \leq 0.6) \\
&\approx 1 - \Phi(0.6) \\
&= 1 - 0.7257 = 0.2743.
\end{aligned}$$

Question 4.2.2. ([4], No. 82) An insurance company issues 1,250 vision care insurance policies. The number of claims filed by a policyholder under a vision care insurance policy during one year is a Poisson random variable with mean 2. Assume the numbers of claims filed by distinct policyholder are independent of one another.

What is the approximate probability that there is a total of between 2,450 and 2,600 claims during a one-year period?

SOLUTION. Let $X_1, X_2, \cdots, X_{1{,}250}$ be the numbers of claims filed by 1,250 policyholders, and let $S = \sum_{k=1}^{1{,}250} X_k$. The question is to find $P(2,450 < S < 2,600)$. Since, for all $k = 1, 2, \cdots, 1,250$, $X_k \sim P(2)$, we have $E[X_k] = Var(X_k) = 2$.

Thus we have

$$\begin{aligned} E[S] &= \sum_{k=1}^{1{,}250} E[X_k] = \sum_{k=1}^{1{,}250} 2 \\ &= 1,250 \times 2 = 2,500. \end{aligned}$$

and, from the condition that $X_1, X_2, \cdots, X_{1{,}250}$ are independent, we see

$$\begin{aligned} Var(S) &= \sum_{k=1}^{1{,}250} Var(X_k) = \sum_{k=1}^{1{,}250} 2 \\ &= 1,250 \times 2 = 50^2. \end{aligned}$$

By the central limit theorem, we see

$$Z \equiv \frac{S - 2,500}{50} \sim N(0, 1^2) \quad \text{approximately}$$

Using the standard normal table, we see that

$$\begin{aligned}
&P(2,450 < S < 2,600)\\
&= P\left(\frac{2,450 - 2,500}{50} < \frac{S - 2,500}{50} < \frac{2,600 - 2,500}{50}\right)\\
&= P(-1 < Z < 2)\\
&= P(Z < 2) - P(Z < -1)\\
&\approx \Phi(2) - \Phi(-1)\\
&= \Phi(2) + \Phi(1) - 1\\
&= 0.9773 + 0.8413 - 1\\
&= 0.8186.
\end{aligned}$$

Question 4.2.3. ([4], No. 84) Let X and Y be the number of hours that a randomly selected person watches movies and sporting events, respectively, during a three-month period. The following information is known about X and Y:

$$\begin{aligned}
E[X] &= 50\\
E[Y] &= 20\\
Var(X) &= 50\\
Var(Y) &= 30\\
Cov(X, Y) &= 10.
\end{aligned}$$

One hundred people are randomly selected and observed for these months. Let T be the total number of hours that these one hundred people watch movies or sporting events during this three-month period.

Approximate the value of $P(T < 7,100)$

Solution. Let X_i and Y_i be the number of hours that a randomly selected i-th person watches movies and sporting events, respectively, for

$i = 1, 2, \cdots, 100$. Since $T = \sum_{i=1}^{100}(X_i + Y_i)$,

$$E[T] = \sum_{i=1}^{100}(E[X_i] + E[Y_i]) = 100 \times (50 + 20) = 7,000.$$

and

$$\begin{aligned} Var(T) &= \sum_{i=1}^{100} Var(X_i + Y_i) \\ &= \sum_{i=1}^{100}(Var(X_i) + Var(Y_i) + 2Cov(X_i, Y_i)) \\ &= 100 \times (50 + 30 + 2 \times 10) \\ &= 100^2 \end{aligned}$$

By the central limit theorem, we see

$$\frac{T - 7,000}{100} \sim N(0, 1^2) \quad \text{approximately} \quad .$$

Using the standard normal table, we see that

$$\begin{aligned} P(T < 7,100) &= P\left(\frac{T - 7,000}{100} < \frac{7,100 - 7,000}{100}\right) \\ &= P\left(\frac{T - 7,000}{100} < 1\right) \\ &\approx \Phi(1) \\ &= 0.8413. \end{aligned}$$

Question 4.2.4. ([4], No. 86) A city has just added 100 new female recruits to its police force. The city will provide a pension to each new hire who remains with the force until retirement. In addition, if the new hire is married at the time of her retirement, a second pension will be provided for her husband. A consulting actuary makes the following assumptions:

(i) Each new recruit has a 0.4 probability of remaining with the police force until retirement.

(ii) Given that a new recruit reaches retirement with the police force, the probability that she is not married at time of retirement is 0.25.

(iii) The number of pensions that the city will provide on behalf of each new hire is independent of the number of pensions it will provide on behalf of any other new hire.

Determine the probability that the city will provide at most 90 pensions to the 100 new hire and their husbands.

SOLUTION. Let N_i denote the number of pensions for i-th new recruit, $i = 1, 2, \cdots, 100$, and let $N = \sum_{i=1}^{100} N_i$. The question is to determine $P(N \le 90)$. Since N is an integer valued random variable, we use the continuous correction (see Remark 2.4.12). From the assumption, we know that

$$N_i = \begin{cases} 0, & \text{w.p.} \quad 1 - 0.4 = 0.6 \\ 1, & \text{w.p.} \quad 0.4 \times 0.25 = 0.1 \\ 2, & \text{w.p.} \quad 0.4 \times 0.75 = 0.3 \end{cases}$$

so that

$$E[N] = \sum_{i=1}^{100} E[N_i] = 100\{(0)(0.6) + (1)(0.1) + (2)(0.3)\} = 70$$

and

$$\begin{aligned} Var(N) &= \sum_{i=1}^{100} Var(N_i) \\ &= 100\{(0)^2(0.6) + (1)^2(0.1) + (2)^2(0.3) - (0.7)^2\} = 81 = 9^2. \end{aligned}$$

Thus, by the standard normal table, we have

$$\begin{aligned} P(N \le 90) &= P(N < 90.5) \\ &= P\left(\frac{N-70}{9} < \frac{90.5-70}{9}\right) \\ &= P\left(\frac{N-70}{9} < 2.28\right) \\ &\approx \Phi(2.28) \\ &= 0.9887. \end{aligned}$$

Question 4.2.5. ([4], No. 87) In any analysis of health care data, ages have been rounded to the nearest multiple of 5 years. The difference between the true age and the rounded age is assumed to be uniformly distributed on the interval from -2.5 year to 2.5 years. The health care data are based on a random sample of 48 people.

What is the approximate probability that the sample mean of the rounded age is within 0.25 years of the sample mean of the true ages?

SOLUTION. Let T_i and R_i denote the true age and rounded age of i-th person in a sample, $i = 1, 2, \cdots, 48$, and let

$$\overline{T} = \frac{1}{48}\sum_{i=1}^{48} T_i \quad \text{and} \quad \overline{R} = \frac{1}{48}\sum_{i=1}^{48} R_i.$$

The question is to find $P(|\overline{T} - \overline{R}| < 0.25)$. Since, $T_i - R_i \sim U(-2.5, 2.5)$ for all $i = 1, 2, \cdots, 48$, we see

$$E[T_i - R_i] = \frac{-2.5 + 2.5}{2} = 0$$

and

$$Var(T_i - R_i) = \frac{(2.5 + 2.5)^2}{12} = \frac{25}{12} = 2.083.$$

Thus we have

$$\begin{aligned} E[\overline{T} - \overline{R}] &= E\left[\frac{1}{48}\sum_{i=1}^{48} T_i - \frac{1}{48}\sum_{i=1}^{48} R_i\right] \\ &= \frac{1}{48}\sum_{i=1}^{48} E[T_i - R_i] \\ &= \left(\frac{1}{48}\right)(48)(0) \\ &= 0 \end{aligned}$$

and

$$\begin{aligned}
Var(\overline{T}-\overline{R}) &= Var\left(\frac{1}{48}\sum_{i=1}^{48} T_i - \frac{1}{48}\sum_{i=1}^{48} R_i\right) \\
&= \left(\frac{1}{48}\right)^2 \sum_{i=1}^{48} Var(T_i - R_i) \\
&= \left(\frac{1}{48}\right)^2 (48)(2.083) \\
&= 0.4339 = 0.2083^2.
\end{aligned}$$

By the central limit theorem,

$$\overline{T}-\overline{R} = \frac{1}{48}\sum_{i=1}^{48}(T_i - R_i) \sim N(0, 0.2083^2) \quad \text{approximately.}$$

Using the standard normal table, we have

$$\begin{aligned}
P(|\overline{T}-\overline{R}| < 0.25) &= P(-0.25 < \overline{T}-\overline{R} < 0.25) \\
&= P\left(\frac{-0.25}{0.2083} < \frac{\overline{T}-\overline{R}}{0.2083} < \frac{0.25}{0.2083}\right) \\
&= P\left(-1.2 < \frac{\overline{T}-\overline{R}}{0.2083} < 1.2\right) \\
&\approx \Phi(1.2) - \Phi(-1.2) \\
&= 2\Phi(1.2) - 1 \\
&= 2 \times 0.8849 - 1 = 0.77.
\end{aligned}$$

CHAPTER 5

Applications to Loss Distributions and Insurance

5.1. How insurance works?

Question 5.1.1. ([4], No. 85) The total claim amount for a health insurance policy follows a distribution with density function

$$f(x) = \frac{1}{1,000} e^{(x/1,000)}, \quad \text{for} \quad x > 0.$$

The premium for the policy is set 100 over the expected total claim amount. If 100 policies are sold, what is the approximate probability that the insurance company will have claims exceeding the premiums collected?

SOLUTION. Let X_i and Q_i denote the total claim amount and the premium for i-th policyholder, respectively, $i = 1, 2, \cdots, 100$, and let $S = \sum_{i=1}^{100} X_i$ and $Q = \sum_{i=1}^{100} Q_i$. The question is to find $P(S \geq Q)$. We are given that for all $i = 1, 2, \cdots, 100$, $X_i \sim E(1/1000)$ and $Q_i = 100 + E[X_i]$. From these we get $E[X_i] = 1,000$ and $Var(X_i) = 1,000^2$ and hence

$$E[S] = \sum_{i=1}^{100} E[X_i] = 100 \times 1,000 = 100,000,$$

$$Var(S) = \sum_{i=1}^{100} Var(X_i) = 100 \times 1,000^2 = 10,000^2$$

and

$$Q = \sum_{i=1}^{100} Q_i = \sum_{i=1}^{100} (100 + E[X_i]) = 100(100 + 1,000) = 110,000.$$

Using the central limit theorem and standard normal table, we get

$$\begin{aligned} P(S \geq Q) &= P(S \geq 110,000) \\ &= P\left(\frac{S - 100,000}{10,000} \geq \frac{110,000 - 100,000}{10,000}\right) \\ &= P\left(\frac{S - 100,000}{10,000} \geq 1\right) \\ &= 1 - P\left(\frac{S - 100,000}{10,000} < 1\right) \\ &\approx 1 - \Phi(1) \\ &= 1 - 0.841 = 0.159. \end{aligned}$$

Question 5.1.2. ([4], No. 113) Two life insurance policies, each with a death benefit of 10,000 and a one-time premium of 500, are sold to a couple, one for each person. The policies will expire at the end of the tenth year. The probability that only the wife will survive at least ten years is 0.025, the probability that only the husband will survive at least ten years is 0.01, and the probability that both of them will survive at least ten years is 0.96.

What is the expected excess of premiums over claims, given that the husband survive at least ten years?

SOLUTION. Let

W = event that the wife survives at least ten years

H = event that the husband survives at least ten years

B = death benefit paid by company to the husband,

and let

$$\begin{aligned}
P &= \text{excess of premiums over benefits}\\
&= (\text{total premiums}) - (\text{death benefit paid by company to the husband})\\
&= 2 \times 500 - B\\
&= 1,000 - B.
\end{aligned}$$

The question is to find $E[P|H]$. We are given that

$$B = \begin{cases} 0, & \text{if} \quad W \quad \text{occurs} \\ 10,000, & \text{if} \quad W^c \quad \text{occurs.} \end{cases}$$

Then

$$\begin{aligned}
E[P|H] &= E[1,000 - B|H]\\
&= 1,000 - E[B|H]\\
&= 1,000 - \{0 \times P(W|H) + 10,000 \times P(W^c|H)\}\\
&= 1,000 - 10,000 \times P(W^c|H).
\end{aligned}$$

Here

$$\begin{aligned}
P(W^c|H) &= \frac{P(W^c \cap H)}{P(H)}\\
&= \frac{P(W^c \cap H)}{P(H \cap [W \cup W^c])}\\
&= \frac{P(W^c \cap H)}{P(H \cap W) + P(H \cap W^c)}\\
&= \frac{0.01}{0.96 + 0.01} = 0.0103.
\end{aligned}$$

Thus we have

$$E[P|H] = 1,000 - 10,000 \times 0.0103 = 1,000 - 103 = 897.$$

5.2. Models for describing a loss random variable

5.3. Modeling the aggregate claims

Question 5.3.1. ([4], No. 123) You are given the following information about N, the annual number of claims for a randomly selected insured:

$$P(N=0) = \frac{1}{2}$$
$$P(N=1) = \frac{1}{3}$$
$$P(N>1) = \frac{1}{6}.$$

Let S denote the total annual claim amount for an insured. When $N=1$, S is exponentially distributed with mean 5. When $N>1$, S is exponentially distributed with mean 8.

Determine $P(4 < S < 8)$.

SOLUTION. We remember that if $X \sim E(\lambda)$, i.e., X is exponentially distributed with mean $1/\lambda$, then the distribution function is of the form $F_X(x) = 1 - e^{-\lambda x}$.

Similarly, by the assumptions, we are given

$$F_{S|N}(x|N=1) = 1 - e^{-\frac{x}{5}}$$

and

$$F_{S|N}(x|N>1) = 1 - e^{-\frac{x}{8}}.$$

Thus

$$
\begin{aligned}
P(4 < S < 8) &= P(4 < S < 8|N = 1)P(N = 1) \\
&\quad + P(4 < S < 8|N > 1)P(N > 1) \\
&= [P(S < 8|N = 1) - P(S < 4|N = 1)]P(N = 1) \\
&\quad + [P(S < 8|N > 1) - P(S < 4|N > 1)]P(N > 1) \\
&= [F_{S|N}(8|N = 1) - F_{S|N}(4|N = 1)]P(N = 1) \\
&\quad + [F_{S|N}(8|N > 1) - F_{S|N}(4|N > 1)]P(N > 1) \\
&= \left[\left(1 - e^{-\frac{8}{5}}\right) - \left(1 - e^{-\frac{4}{5}}\right)\right] \times \frac{1}{3} \\
&\quad + \left[\left(1 - e^{-\frac{8}{8}}\right) - \left(1 - e^{-\frac{4}{8}}\right)\right] \times \frac{1}{6} \\
&= \frac{1}{3}\left(e^{-\frac{4}{5}} - e^{-\frac{8}{5}}\right) + \frac{1}{6}\left(e^{-\frac{4}{8}} - e^{-\frac{8}{8}}\right) \\
&= 0.122.
\end{aligned}
$$

5.4. The effect of coverage adjustments on a loss distribution

Question 5.4.1. ([4], No. 40) An insurance policy pays for a random loss X subject to a deductible of C, where $0 < C < 1$. The loss amount is modeled as a continuous random variable with density function

$$
f_X(x) = \begin{cases} 2x, & \text{for } 0 < x < 1 \\ 0, & \text{otherwise.} \end{cases}
$$

Given a random loss X, the probability that the insurance payment is less than 0.5 is equal to 0.64.

Calculate C.

SOLUTION. Let Y be the insurance payment for a random loss X. Then

$$Y = \begin{cases} 0, & \text{if } 0 < X \leq C \\ X - C, & \text{if } C < X < 1. \end{cases}$$

Since $P(Y < 0.5) = 0.64$, we have

$$\begin{aligned} 0.8^2 = 0.64 = P(Y < 0.5) &= P(0 < X - C < 0.5) \\ &= P(0 < X < C + 0.5) \\ &= \int_0^{C+0.5} 2x dx = [x^2]_0^{C+0.5} \\ &= (C + 0.5)^2 \end{aligned}$$

so that $C = -0.5 \pm 0.8$. Since $0 < C < 1$, we get $C = 0.3$.

Question 5.4.2. ([4], No. 51) A manufacturer's annual losses follow a distribution with density function

$$f(x) = \begin{cases} \frac{2.5(0.6)^{2.5}}{x^{3.5}}, & \text{for } x > 0.6 \\ 0, & \text{otherwise.} \end{cases}$$

To cover its losses, the manufacturer purchases an insurance policy with an annual deductible of 2.

What is the mean of the manufacturer' annual losses not paid by the insurance policy?

SOLUTION. Let X denote the manufacturer's annual loss. Then X is represented by $X = Y_l + Y_s$, where Y_l is a claim amount per loss and Y_s is a retained loss (see Remark 5.4.1).

The question is to find $E[Y_s]$. Since

$$Y_l = \begin{cases} 0, & \text{if } 0.6 < X < 2 \\ X - 2, & \text{if } X \geq 2, \end{cases}$$

we have

$$Y_s = g(X) = \begin{cases} X, & \text{if } 0.6 < X < 2 \\ 2, & \text{if } X \geq 2. \end{cases}$$

Thus

$$\begin{aligned} E[Y_s] &= E[g(X)] \\ &= \int_{0.6}^{2} g(x) f_X(x) dx + \int_{2}^{\infty} g(x) f_X(x) dx \\ &= \int_{0.6}^{2} x \left(\frac{2.5(0.6)^{2.5}}{x^{3.5}} \right) dx + \int_{2}^{\infty} 2 \left(\frac{2.5(0.6)^{2.5}}{x^{3.5}} \right) dx \\ &= \left[-\frac{2.5(0.6)^{2.5}}{1.5x^{1.5}} \right]_{0.6}^{2} + \left[-\frac{2(0.6)^{2.5}}{x^{2.5}} \right]_{2}^{\infty} \\ &= 0.9343. \end{aligned}$$

Question 5.4.3. ([4], No. 52) An insurance company sells a one-year automobile policy with a deductible of 2. The probability that the insured will incur a loss is 0.05. If there is a loss, the probability of a loss of amount N is K/N, for $N = 1, 2, \cdots, 5$ and K a constant. These are the only possible loss amounts and no more than one loss can occur.

Determine the net premium for this policy.

SOLUTION. Let Y be the claim amount paid per loss. The question is to determine $E[Y]$. Because of the deductible of 2, Y is defined by

$$Y = g(N) = \begin{cases} 0, & \text{if } N \leq 2 \\ N - 2, & \text{if } N > 2. \end{cases}$$

To find $E[Y]$, we need to determine the pmf $p_N(n)$ of N. From

$$\begin{aligned} 1 &= \sum_{n=1}^{5} P(N = n | \text{a loss occurs}) \\ &= \frac{K}{1} + \frac{K}{2} + \frac{K}{3} + \frac{K}{4} + \frac{K}{5} \\ &= \frac{137K}{60}, \end{aligned}$$

we get $K = 60/137$. Thus we have

$$\begin{aligned}
p_N(n) &= P(N = n) \\
&= P(N = n|\text{a loss occurs})P(\text{a loss occurs}) \\
&\quad + P(N = n|\text{no loss occurs})P(\text{no loss occurs}) \\
&= \frac{K}{n} \times 0.05 + 0 \times 0.95 \\
&= \frac{K}{20n} \\
&= \frac{60}{(137)(20)n} = \frac{3}{137n}.
\end{aligned}$$

Therefore

$$\begin{aligned}
E[Y] = E[g(N)] &= \sum_{n=1}^{5} g(n)p_N(n) \\
&= g(1)p_N(1) + g(2)p_N(2) + g(3)p_N(3) + g(4)p_N(4) + g(5)p_N(5) \\
&= (0)\left(\frac{3}{137 \times 1}\right) + (0)\left(\frac{3}{137 \times 2}\right) + (3-2)\left(\frac{3}{137 \times 3}\right) \\
&\quad + (4-2)\left(\frac{3}{137 \times 4}\right) + (5-2)\left(\frac{3}{137 \times 5}\right) \\
&= 0.0314.
\end{aligned}$$

Question 5.4.4. ([4], No. 53) An insurance policy reimburses a loss up to a benefit limit of 10. The policyholder's loss, X, follows a distribution with density function

$$f_X(x) = \begin{cases} \frac{2}{x^3}, & \text{for } x > 1 \\ 0, & \text{otherwise.} \end{cases}$$

What is the expected value of the benefit paid under the insurance policy?

SOLUTION. Let Y_l denote the benefit paid for the policyholder's loss. The question is to find $E[Y_l]$. By assumption, Y_l is defined by

$$Y_l = g(X) = \begin{cases} X, & \text{if } 1 < X \le 10 \\ 10, & \text{if } X > 10. \end{cases}$$

Then

$$\begin{aligned} E[Y_l] = E[g(X)] &= \int_{-\infty}^{\infty} g(x) f_X(x) dx \\ &= \int_1^{10} x \frac{2}{x^3} dx + \int_{10}^{\infty} 10 \frac{2}{x^3} dx \\ &= \left[-\frac{2}{x} \right]_1^{10} + \left[-\frac{10}{x^2} \right]_{10}^{\infty} \\ &= 2 - \frac{2}{10} + \frac{1}{10} - 0 = 1.9. \end{aligned}$$

Alternatively, we can use the following formula (5.4.1).

$$E[Y_l] = 1 + \int_1^{10} (1 - F_X(x)) dx.$$

Since

$$\begin{aligned} F_X(x) &= \int_{-\infty}^{x} f_X(u) du = \int_1^x \frac{2}{u^3} du \\ &= \left[-\frac{1}{u^2} \right]_1^x = 1 - \frac{1}{x^2}, \end{aligned}$$

we have

$$\begin{aligned} E[Y_l] &= 1 + \int_1^{10} (1 - F_X(x)) dx \\ &= 1 + \int_1^{10} \left\{ 1 - \left(1 - \frac{1}{x^2} \right) \right\} \\ &= 1 + \int_1^{10} \frac{1}{x^2} dx \\ &= 1 + \left[-\frac{1}{x} \right]_1^{10} \\ &= 1 + 1 - \frac{1}{10} = 2 - 0.1 = 1.9. \end{aligned}$$

Question 5.4.5. ([4], No. 56) An insurance policy is written to cover a loss, X, where X has a uniform distribution on [0, 1,000].

At what level must a deductible be set in order for the expected payment to be 25% of what it would be with no deductible?

SOLUTION. Let Y be the claim amount paid per loss subject to a deductible D. That is,

$$Y = g(X) = \begin{cases} 0, & \text{if } X \leq D \\ X - D, & \text{if } X > D. \end{cases}$$

We are given that

$$E[Y] = 0.25E[X]$$

The question is to determine D. Since $X \sim U(0, 1,000)$, $f_X(x) = 1/1,000$ and $E[X] = 1,000/2 = 500$. Thus $E[Y] = (0.25)(500) = 125$. On the other hand,

$$\begin{aligned} E[Y] = E[g(X)] &= \int_D^{1,000} g(x) f_X(x) dx \\ &= \int_D^{1,000} (x - D) \frac{1}{1,000} dx \\ &= \left[\frac{1}{1,000} \frac{(x-D)^2}{2} \right]_D^{1,000} = \frac{(1,000 - D)^2}{2,000}. \end{aligned}$$

Hence we have

$$125 = \frac{(1,000 - D)^2}{2,000},$$

i.e.,

$$(1,000 - D)^2 = 125 \times 2,000 = 500^2.$$

From this, we get $D = 500$ or 1,500. Since $D < 1,000$, the answer is $D = 500$.

Question 5.4.6. ([4], No. 65) The owner of an automobile insures it against damage by purchasing an insurance policy with a deductible of 250.

In the event that the automobile is damaged, repair costs can be modeled by a uniform random variable on the interval (0, 1,500).

Determine the standard deviation of the insurance payment in the event that the automobile is damaged.

SOLUTION. Let X and Y be repair costs and insurance payment, respectively, in the event that the automobile is damaged. The question is to find $\sqrt{Var(Y)}$. We are given that $X \sim U(0, 1,500)$, i.e., $f_X(x) = 1/1,500$, and

$$Y = g(X) = \begin{cases} 0, & \text{if } X \leq 250 \\ X - 250, & \text{if } X > 250. \end{cases}$$

Then we have

$$\begin{aligned} E[Y] = E[g(X)] &= \int_{-\infty}^{\infty} g(x) f_X(x) dx \\ &= \int_{250}^{1,500} (x - 250) \frac{1}{1,500} dx \\ &= \left[\frac{1}{1,500} \frac{(x - 250)^2}{2} \right]_{250}^{1,500} = 521. \end{aligned}$$

and

$$\begin{aligned} E[Y^2] = E[g(X)] &= \int_{-\infty}^{\infty} g(x)^2 f_X(x) dx \\ &= \int_{250}^{1,500} (x - 250)^2 \frac{1}{1,500} dx \\ &= \left[\frac{1}{1,500} \frac{(x - 250)^3}{3} \right]_{250}^{1,500} = 434,028. \end{aligned}$$

Therefore we have

$$\sqrt{Var(Y)} = \sqrt{E[X^2] - E[X]^2} = \sqrt{434,028 - 521^2} = 403.$$

Question 5.4.7. ([4], No. 68) An insurance policy reimburses dental expense, X, up to a maximum benefit of 250. The probability density function for X is:

$$f_X(x) = \begin{cases} ce^{-0.004x}, & \text{for } x \geq 0 \\ 0, & \text{otherwise,} \end{cases}$$

where c is a constant.

Calculate the median benefit for this policy.

SOLUTION. Let Y be the claim benefit paid for dental expense X. Since a policy limit 250 is applied, Y is defined by

$$Y = g(X) = \begin{cases} X, & \text{if } X \leq 250 \\ 250, & \text{if } X > 250. \end{cases}$$

The question is to find m_0 satisfying $0.5 = P(Y \leq m_0)$. From

$$1 = \int_{-\infty}^{\infty} f_X(x)dx = \int_0^{\infty} ce^{-0.004x}dx = \frac{c}{0.004},$$

we get $c = 0.004$ and so

$$f_X(x) = \begin{cases} 0.004e^{-0.004x}, & \text{for } x \geq 0 \\ 0, & \text{otherwise,} \end{cases}$$

This means $X \sim E(0.004)$. Then

$$\begin{aligned} 0.5 = P(Y \leq m_0) &= P(g(X) \leq m_0) = P(X \leq g^{-1}(m_0)) \\ &= \int_0^{g^{-1}(m_0)} f_X(x)dx \\ &= \int_0^{g^{-1}(m_0)} 0.004e^{-0.004x}dx \\ &= 1 - 0.004e^{-0.004g^{-1}(m_0)}, \end{aligned}$$

so that

$$e^{-0.004g^{-1}(m_0)} = 0.5 = 2^{-1}.$$

Taking ln in both side, we have

$$g^{-1}(m_0) = 250 \ln 2 = 173.29,$$

that is,

$$m_0 = g(173.29).$$

Since $173.29 \leq 250$, by the definition of the function g, $m_0 = 173.29$.

Question 5.4.8. ([4], No. 93) A family buys two policies from the same insurance company. Losses under the two policies are independent and have continuous uniform distributions on the interval from 0 to 10. One policy has a deductible of 1 and the other has a deductible of 2. The family experiences exactly one loss under each policy.

Calculate the probability that the total benefit paid to the family does not exceed 5.

SOLUTION. Let X_1 and X_2 be loss amounts covered by the policy having deductible of 1 and 2, respectively, and let

$$Y_1 = \begin{cases} 0, & \text{if } X_1 < 1 \\ X_1 - 1, & \text{if } X_1 \geq 1 \end{cases} \quad \text{and} \quad Y_2 = \begin{cases} 0, & \text{if } X_2 < 2 \\ X_2 - 2, & \text{if } X_2 \geq 2. \end{cases}$$

The question is to calculate $P(Y_1 + Y_2 < 5)$. We are given that $X_1, X_2 \sim U(0, 10)$ and they are independent. Thus the joint pdf of X_1 and X_2 is of the form

$$f_{X_1,X_2}(x_1, x_2) = f_{X_1}(x_1) f_{X_2}(x_2) = \left(\frac{1}{10}\right)\left(\frac{1}{10}\right) = \frac{1}{100}$$

for all $0 < x_1, x_2 < 10$. Let D be the support of $f_{X_1,X_2}(x_1, x_2)$, i.e., $D = \{(x_1, x_2) | 0 < x_1, x_2 < 10\}$. Then

$$\begin{aligned} P(Y_1 + Y_2 < 5) &= P(X_1 - 1 + X_2 - 2 < 5, X_1 < 6, X_2 < 7) \\ &= P(X_1 + X_2 < 8, X_1 < 6, X_2 < 7) \\ &= P((X_1, X_2) \in A), \end{aligned}$$

where

$$
\begin{aligned}
A &= \{(x_1, x_2) \in D | x_1 + x_2 < 8, x_1 < 6, x_2 < 7\} \\
&= \{(x_1, x_2) | 0 < x_1 < 6, 0 < x_2 < 2\} \\
&\quad + \{(x_1, x_2) | 0 < x_1 < 1, 2 < x_2 < 7\} \\
&\quad\quad + \{(x_1, x_2) | 1 < x_1 < 6, 2 < x_2 < 8 - x_1\} \\
&= A_1 + A_2 + A_3.
\end{aligned}
$$

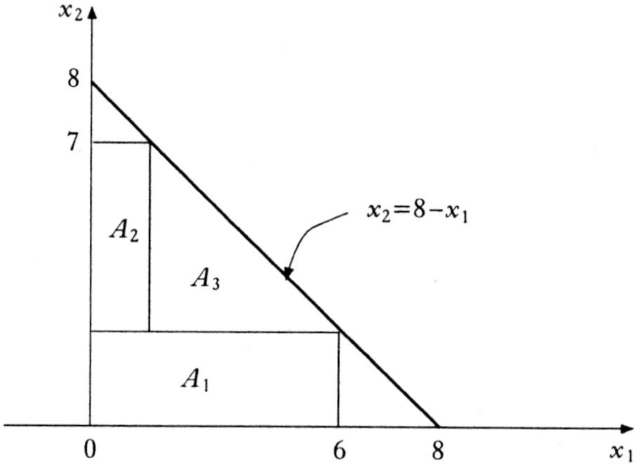

Thus

$$
\begin{aligned}
&P(Y_1 + Y_2 < 5) \\
&= P((X_1, X_2) \in A_1) + P((X_1, X_2) \in A_2) + P((X_1, X_2) \in A_3) \\
&= \sum_{k=1}^{3} \int\int_{A_k} f_{X_1, X_2}(x_1, x_2) dx_1 dx_2 \\
&= \sum_{k=1}^{3} \int\int_{A_k} \frac{1}{100} dx_1 dx_2 \\
&= \frac{1}{100} \{(\text{area of } A_1) + (\text{area of } A_2) + (\text{area of } A_3)\} \\
&= \frac{1}{100} \{(6)(2) + (1)(5) + (1/2)(5)(5)\} \\
&= 0.295.
\end{aligned}
$$

Table of Standard Normal Distribution

The table represents the values of $\alpha = \Phi(z) = P(Z \leq z) = \int_{-\infty}^{z} exp\{-\frac{x^2}{2}\}dx$ for any $z \in [0, 3.09]$. Conversely for any $\alpha \in (0, 1)$, we can find z.

z	0.00	0.01	0.02	0.03	0.04	0.05	0.06	0.07	0.08	0.09
0.0	0.5000	0.5040	0.5080	0.5120	0.5160	0.5199	0.5239	0.5279	0.5319	0.5359
0.1	0.5398	0.5438	0.5478	0.5517	0.5557	0.5596	0.5636	0.5675	0.5714	0.5753
0.2	0.5793	0.5832	0.5871	0.5910	0.5948	0.5987	0.6026	0.6064	0.6103	0.6141
0.3	0.6179	0.6217	0.6255	0.6293	0.6331	0.6368	0.6406	0.6443	0.6480	0.6517
0.4	0.6554	0.6591	0.6628	0.6664	0.6700	0.6736	0.6772	0.6808	0.6844	0.6879
0.5	0.6915	0.6950	0.6985	0.7019	0.7054	0.7088	0.7123	0.7157	0.7190	0.7224
0.6	0.7257	0.7291	0.7324	0.7357	0.7389	0.7422	0.7454	0.7486	0.7517	0.7549
0.7	0.7580	0.7611	0.7642	0.7673	0.7704	0.7734	0.7764	0.7794	0.7823	0.7852
0.8	0.7881	0.7910	0.7939	0.7967	0.7995	0.8023	0.8051	0.8078	0.8106	0.8133
0.9	0.8159	0.8186	0.8212	0.8238	0.8264	0.8289	0.8315	0.8340	0.8365	0.8389
1.0	0.8413	0.8438	0.8461	0.8485	0.8508	0.8531	0.8554	0.8577	0.8599	0.8621
1.1	0.8643	0.8665	0.8686	0.8708	0.8729	0.8749	0.8770	0.8790	0.8810	0.8830
1.2	0.8849	0.8869	0.8888	0.8907	0.8925	0.8944	0.8962	0.8980	0.8997	0.9015
1.3	0.9032	0.9049	0.9066	0.9082	0.9099	0.9115	0.9131	0.9147	0.9162	0.9177
1.4	0.9192	0.9207	0.9222	0.9236	0.9251	0.9265	0.9279	0.9292	0.9306	0.9319
1.5	0.9332	0.9345	0.9357	0.9370	0.9382	0.9394	0.9406	0.9418	0.9429	0.9441
1.6	0.9452	0.9463	0.9474	0.9484	0.9495	0.9505	0.9515	0.9525	0.9535	0.9545
1.7	0.9554	0.9564	0.9573	0.9582	0.9591	0.9599	0.9608	0.9616	0.9625	0.9633
1.8	0.9641	0.9649	0.9656	0.9664	0.9671	0.9678	0.9686	0.9693	0.9699	0.9706
1.9	0.9713	0.9719	0.9726	0.9732	0.9738	0.9744	0.9750	0.9756	0.9761	0.9767
2.0	0.9772	0.9778	0.9783	0.9788	0.9793	0.9798	0.9803	0.9808	0.9812	0.9817
2.1	0.9821	0.9826	0.9830	0.9834	0.9838	0.9842	0.9846	0.9850	0.9854	0.9857
2.2	0.9861	0.9864	0.9868	0.9871	0.9875	0.9878	0.9881	0.9884	0.9887	0.9890
2.3	0.9893	0.9896	0.9898	0.9901	0.9904	0.9906	0.9909	0.9911	0.9913	0.9916
2.4	0.9918	0.9920	0.9922	0.9925	0.9927	0.9929	0.9931	0.9932	0.9934	0.9936
2.5	0.9938	0.9940	0.9941	0.9943	0.9945	0.9946	0.9948	0.9949	0.9951	0.9952
2.6	0.9953	0.9955	0.9956	0.9957	0.9959	0.9960	0.9961	0.9962	0.9963	0.9964
2.7	0.9965	0.9966	0.9967	0.9968	0.9969	0.9970	0.9971	0.9972	0.9973	0.9974
2.8	0.9974	0.9975	0.9976	0.9977	0.9977	0.9978	0.9979	0.9979	0.9980	0.9981
2.9	0.9981	0.9982	0.9982	0.9983	0.9984	0.9984	0.9985	0.9985	0.9986	0.9986
3.0	0.9987	0.9987	0.9987	0.9988	0.9988	0.9989	0.9989	0.9989	0.9990	0.9990

Examples: $\Phi(1.96) = 0.975$. Since the pdf is symmetric w.r.t. 0, we can find

$\Phi(-1.96) = 0.025$

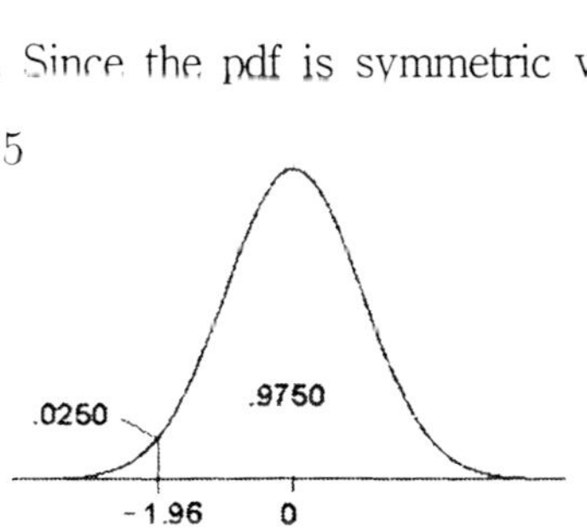

Bibliography

[1] N.L. Bowers Jr., H.U. Gerber, J.C. Hickman, D.A. Jones and C.J. Nesbitt, Accuarial Mathematics, The Society of Actuaries, Itasca, Illinois, 1986.

[2] S.A. Broverman, Study Manual - SOA Course 1 Examination and CAS Exam 1, ACTEX Publications, Inc., 2001.

[3] SOA/CAS, SOA/CAS Course 1 Exam - May 2000.

[4] SOA/CAS Exam P/1 Sample Questions P-09-05, http://www.soa.org/Exam & Jobs/Spring 2007 Basic Education Catalog, 2007.

Index